HUTCHINSON POCKET

Dictionary of
Mathematics

HUTCHINSON POCKET

Dictionary of
Mathematics

Helicon

Copyright © Helicon Publishing Ltd. 1993

All rights reserved

Helicon Publishing Ltd
42 Hythe Bridge Street
Oxford OX1 2EP

Typeset by Roger Walker Graphic Design and Typography,
Maidenhead, Berkshire

Printed and bound in Great Britain by
Unwin Brothers Ltd, Old Woking, Surrey

ISBN 0 09 178107 8

British Cataloguing in Publication Data

A catalogue record for this book is available
from the British Library

Editorial director
Michael Upshall

Consultant editor
Peter Kaner

Project editor
Sara Jenkins-Jones

Text editor
Catherine Thompson

Art editor
Terence Caven

Additional page make-up
Helen Bird

Production
Tony Ballsdon

A

abacus ancient calculating device made up of a frame of parallel wires on which beads are strung.

The wires define place value (for example, in the decimal number system each successive wire, counting from right to left, would stand for units, tens, hundreds, thousands, and so on) and beads are slid to the top of each wire in order to represent the digits of a particular number. On a simple decimal abacus, for example, the number 8,493 would be entered by sliding three beads on the first wire (three units), nine beads on the second wire (nine tens), four beads on the third wire (four hundreds), and eight beads on the fourth wire (eight thousands).

abscissa in ◊coordinate geometry, the x-coordinate of a point – that is, the horizontal distance of that point from the vertical or y-axis. For example, a point with the coordinates (4, 3) has an abscissa of 3.

The y-coordinate of a point is known as the ◊ordinate.

abscissa

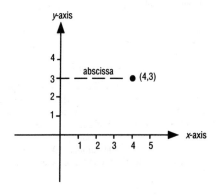

absolute value or *modulus* the value, or magnitude, of a number irrespective of its sign. The absolute value of a number n is written $|n|$ (or sometimes as mod n), and is defined as the positive square root of n^2. For example, the numbers -5 and 5 have the same absolute value: $|5| = |-5| = 5$.

acceleration the rate of change of the velocity of a moving body. It is usually measured in metres per second per second (m s^{-2}). If a body travelling in a straight line accelerates at, for example, 5 m s^{-2}, its velocity may be said to be increasing by 5 m s^{-1} every second. The acceleration of a car is usually judged by the time taken by the car to reach 60 mph from zero. The fewer the seconds the greater the acceleration.

The average acceleration a of an object travelling in a straight line over a period of time t may be calculated using the formula:

$$a = \frac{\text{change of velocity}}{t}$$

A negative answer shows that the object is slowing down (decelerating).

Since velocity is a ⊳vector quantity (possessing both magnitude and direction), it follows that acceleration is also a vector. The acceleration vector of an object following a curved path is, therefore, the change in its velocity vector per unit time.

Acceleration due to gravity (g) is the acceleration of a body falling freely under the influence of the Earth's gravitational field; it varies slightly according to position on the Earth's surface, but is considered internationally to have the value 9.806 m s^{-2}.

accuracy measure of the precision of a number. The degree of accuracy depends on how many figures or decimal places are used in ⊳rounding off the number. For example, the result of a calculation or measurement (such as 13.429314 s) might be rounded off to three decimal places (13.429 s), to two decimal places (13.43 s), to one decimal place (13.4 s), or to the nearest whole number (13 s). The first answer is more accurate than the second, the second more accurate than the third, and so on.

Alternatively, a result might be presented to a certain number of ◊significant figures (digits that are important because of their place value). For example, the number 409,318 might be expressed to an accuracy of four significant figures (409,300), three significant figures (409,000), two significant figures (410,000), or one significant figure (400,000). Here again, the first answer is more accurate than the second, and so on.

acre traditional English land measure equal to 4,840 square yards (4,047 sq m/0.405 ha). Originally meaning a field, it was the area that a pair of oxen could plough in a day.

acute angle between 0° and 90°; that is, a turn that is less than a quarter of a circle.

addition the operation of combining two numbers to form a sum; thus, 7 + 4 = 11. It is one of the four basic operations of arithmetic (the others are subtraction, multiplication, and division).

adjacent angles a pair of angles meeting at a common vertex (corner) and sharing a common arm. Two adjacent angles lying on the same side of a straight line add up to 180°.

adjacent side in a right-angled triangle, the side that is next to a given angle but is not the hypotenuse (the side opposite the right angle). The third side is the side opposite to the given angle.

algebra branch of mathematics in which the general properties of numbers are studied by using symbols, usually letters, to represent variables and unknown quantities. Algebra often involves the use and rearranging of ◊equations. For example, the algebraic statement

$$(x + y)^2 = x^2 + 2xy + y^2$$

is true for all values of x and y. If $x = 7$ and $y = 3$, for instance,

$$(7 + 3)^2 = 7^2 + 2(7 \times 3) + 3^2 = 100.$$

Algebra is used in many areas of mathematics – for example, matrix algebra and Boolean algebra (the latter is used in working out the logic for computers).

adjacent angles

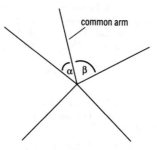

common arm

α β

α and β are adjacent angles

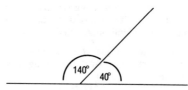

140° 40°

adjacent angles on a straight line add up to 180° and
are therefore said to be supplementary

algebraic fraction fraction in which letters are used to represent
numbers – for example,

$$\frac{a}{b}, \frac{xy}{z^2} \text{ and } \frac{1}{(x+y)}$$

As with numerical fractions, algebraic fractions may be simplified
or factorized. Two equivalent algebraic fractions can be cross-multiplied; for example, if

$$\frac{a}{b} = \frac{c}{d}$$

then

$$ad = bc$$

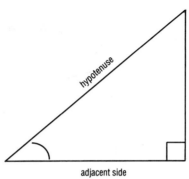

adjacent side

adjacent side

(In the same way, the two equivalent numerical fractions $^2/_3$ and $^4/_6$ can be cross-multiplied to give cross-products that are both 12).

algorithm procedure or series of steps that can be used to solve a problem. A ◊flow chart is a visual representation of an algorithm.

alternate angles a pair of angles that lie on opposite sides and at opposite ends of a transversal (a line that cuts two or more lines in the same plane). The alternate angles formed by a transversal of two parallel lines are equal.

altitude in geometry, the perpendicular distance from a ◊vertex (corner) of a figure, such as a triangle, to the base (the side opposite the vertex).

AND rule rule used for the finding the combined probability of two or more independent events occurring at the same time (or one after the other). If two events E_1 and E_2 are independent (have no effect on each other) and the probabilities of their taking place are p_1 and p_2, respectively, then the combined probability p that both E_1 *and* E_2 will happen is given by

$$p = p_1 \times p_2$$

alternate angles

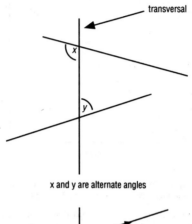

x and y are alternate angles

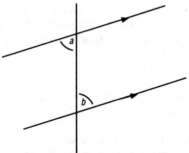

where a transversal cuts through a pair of parallel
lines the alternate angles *a* and *b* are equal

For example, if a blue die and a red die are thrown together, the probability of a blue six is 1/6, and the probability of a red six is 1/6. Therefore, the probability of both a red six and a blue six being thrown is

altitude

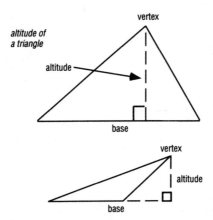

altitude of a triangle

vertex

altitude

base

vertex

altitude

base

two altitudes of a quadrilateral

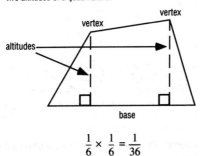

vertex

vertex

altitudes

base

$$\frac{1}{6} \times \frac{1}{6} = \frac{1}{36}$$

By contrast, the ◊OR rule is used for finding the probability of either one event or another taking place.

angle a pair of rays (half-lines) that share a common endpoint but do not lie on the same line. Angles are measured in ◊degrees (°) or

angle

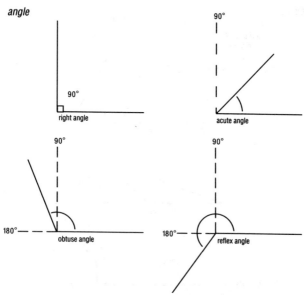

Oradians, and are classified generally by their degree measures. *Acute angles* are less than 90°; *right angles* are exactly 90°; *obtuse angles* are greater than 90° but less than 180°; *reflex angles* are greater than 180° but less than 360°. Angles can be measured by using a protractor.

No angle is classified as having a measure of 180°, as by definition such an 'angle' is actually a straight line.

annulus in geometry, the plane area between two concentric circles, making a flat ring.

anticlockwise direction of rotation, opposite to the way the hands of a clock turn.

apex the highest point of a triangle, cone, or pyramid – that is, the vertex (corner) opposite a given base.

approximation rough estimate of a given value. For example, for ◊pi (which has a value of 3.1415926 correct to seven decimal places), 3 is a rough approximation to the nearest whole number.

Arabic numerals or *Hindu–Arabic numerals* the symbols 0, 1, 2, 3, 4, 5, 6, 7, 8, 9, early forms of which came from India. They spread in use through the Arab countries before being adopted by the peoples of Europe during the Middle Ages, in place of ◊Roman numerals.

Unlike Roman numerals, Arabic numerals include a symbol for zero, which allows us to have a ◊place-value system and enables calculations to be made on paper, instead of by ◊abacus.

arc in geometry, a section of a curved line or circle. A circle has three types of arc: a *semicircle*, which is exactly half of the circle; *minor arcs*, which are less than the semicircle; and *major arcs*, which are greater than the semicircle.

The arcs of a circle are measured in degrees, according to the angle formed by joining its two ends to the centre of the circle. A semicircle is therefore 180°, whereas a minor arc will always be less than 180° (obtuse) and a major arc will always be greater than 180° but less than 360° (reflex).

The length of an arc of angle τ may be calculated using the formula

$$\text{arc length} = \frac{\tau}{360} \times 2\pi r$$

where r is the radius of the circle ($2\pi r$ is the circumference of the circle).

arc minute, arc second units for measuring small angles. An arc minute (symbol') is one-sixtieth of a degree, and an arc second (symbol") is one-sixtieth of an arc minute.

area the size of a surface. It is measured in square units, usually square centimetres (cm^2), square metres (m^2), or square kilometres (km^2). ◊Surface area is the area of the outer surface of a solid.

arithmetic branch of mathematics concerned with the study of numbers and their properties. The fundamental operations of arithmetic are

area

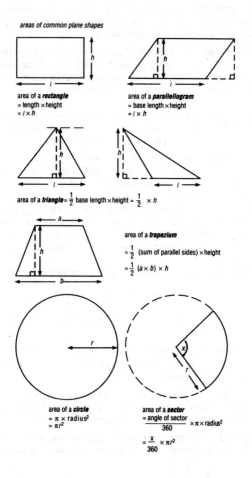

areas of common plane shapes

area of a ***rectangle***
= length × height
= $l \times h$

area of a ***parallellogram***
= base length × height
= $l \times h$

area of a ***triangle*** = $\frac{1}{2}$ base length × height = $\frac{1}{2}$ × h

area of a ***trapezium***

= $\frac{1}{2}$ (sum of parallel sides) × height

= $\frac{1}{2} (a \times b) \times h$

area of a ***circle***
= π × radius²
= πr^2

area of a ***sector***
= $\dfrac{\text{angle of sector}}{360}$ × π × radius²

= $\dfrac{x}{360} \times \pi r^2$

addition, subtraction, multiplication, and division. Raising to ◊powers (for example, squaring or cubing a number), the extraction of roots (such as square roots), percentages, fractions, and ratios are developed from these operations.

Properties of numbers:

associative law All the properties of numbers may be deduced from this law, which states that the sum of a set of numbers is the same whatever the order of addition, and that the product of a set of numbers is the same whatever the order of multiplication.

commutative law A special case of the associative law produces commutativity where there are only two numbers in the set.

$$a + b = b + a$$

$$ab = ba$$

distributive law The distributive law for multiplication over addition states that, given a set of numbers a, b, c, ... and a multiplier m,

$$m(a + b + c + ...) = ma + mb + mc + ...$$

For example,

$$9 \times 132 = (9 \times 100) + (9 \times 30) + (9 \times 2).$$

The distributive law does not apply for addition over multiplication; for example,

$$7 + (3 \times 5) \neq (7 + 3) \times (7 + 5).$$

identities Zero is described as the identity for addition because adding zero to any number has no effect on that number.

$$n + 0 = 0 + n = n$$

One is the identity for multiplication because multiplying any number by one leaves that number unchanged.

$$n \times 1 = 1 \times n = n.$$

negatives Every number has a negative $-n$ such that

$$n + (-n) = 0$$

inverse Every number (except 0) has an inverse $1/n$ such that

$$n \times \frac{1}{n} = 1.$$

These laws can be verified using a calculator.

arithmetic mean the average of a set of n numbers, obtained by adding the numbers and dividing by n. For example, the arithmetic mean of the set of five numbers 1, 3, 6, 8, and 12 is $(1 + 3 + 6 + 8 + 12)/5 = 6$.

The term 'average' is often used to refer only to the arithmetic mean, even though the mean is in fact only one form of average (the others include ◊median and ◊mode).

arithmetic progression or *arithmetic sequence* sequence of numbers or terms that have a common difference between any one term and the next in the sequence. For example, 2, 7, 12, 17, 22, 27, ... is an arithmetic sequence with a common difference of 5.

The nth term in any arithmetic progression can be found using the formula

$$n\text{th term} = a + (n - 1)d$$

where a is the first term and d is the common difference. For example, to find the 7th term of a sequence in which the first term is 2 and the common difference is 3:

$$n = 7, a = 3, \text{ and } d = 3,$$

$$7\text{th term} = 3 + (7 - 1) \times 3 = 21$$

An *arithmetic series* is the sum of the terms in an arithmetic sequence. For a sequence with the first term a and common difference d, the sum of n is:

$$na + \tfrac{1}{2}n(n - 1)d$$

Compare with ◊geometric progression.

array collection of numbers (or letters representing numbers) arranged in rows and columns.

$$a_1 \qquad a_2$$
$$a_3 \qquad a_4$$
$$a_5 \qquad a_6$$

A ◊matrix is an array shown inside a pair of brackets; it indicates that the array should be treated as a single entity.

$$\begin{bmatrix} a_1 & a_2 \\ a_3 & a_4 \\ a_5 & a_6 \end{bmatrix}$$

associative operation an operation in which the outcome is independent of the grouping of the numbers or symbols concerned. For example, addition and multiplication are associative operations.

$$(7 + 2) + 4 = 7 + (2 + 4) = 13$$
$$(4 \times 3) \times 2 = 4 \times (3 \times 2) = 24$$

Subtraction and division are not associative.

$$(9 - 5) - 2 \neq 9 - (5 - 2)$$
$$(12 \div 4) \div 2 \neq 12 \div (4 \div 2)$$

Compare ◊commutative operation and ◊distributive operation.

asymptote in ◊coordinate geometry, a straight line that a curve approaches more and more closely but never reaches. The x and y axes are asymptotes to the graph of $xy =$ constant (a rectangular ◊hyperbola).

average the typical member of a set of data, usually the ◊arithmetic mean. The term is also used to refer to the middle member of the set when ordered in size (the ◊median), and the most commonly occurring item of data (the ◊mode), as in 'the average family'.

axiom a statement that is assumed to be true and upon which theorems are proved by using logical deduction; for example, things that are equal to the same thing are equal to each other, and two straight lines cannot enclose a space. The Greek mathematician Euclid used a series

asymptote

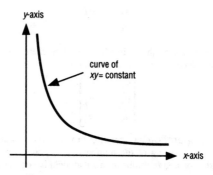

of axioms that he considered could not be demonstrated in terms of simpler concepts to prove his geometrical theorems.

axis (plural *axes*) one of the reference lines by which a point on a graph may be located. A horizontal axis is usually referred to as the *x*-axis, and a vertical axis as the *y*-axis. The term is also used to refer to the imaginary line about which an object may be said to be symmetrical (*axis of symmetry*) – for example, the diagonal of a square – or the line about which an object may revolve (*axis of rotation*).

axis

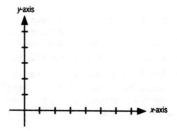

B

bar chart in statistics, a way of displaying data, using horizontal or vertical bars. The heights or lengths of the bars are proportional to the quantities they represent. Bar charts are often represented horizontally and are useful for making comparisons.

bar chart

bar chart showing, for selected countries, the number of cars per 1, 000 of population (1989)

USA	496
West Germany	489
Canada	466
Switzerland	430
Italy	426
Sweden	413
France	410
UK	372
Belgium	366
Spain	290
Japan	263
Greece	128

base the number of different single-digit symbols used in a particular number system. For example, in the usual (decimal) counting system of numbers (with symbols 0, 1, 2, 3, 4, 5, 6, 7, 8, 9) the base is 10; in the ◊binary number system, which has only the symbols 1 and 0, the base is two.

base

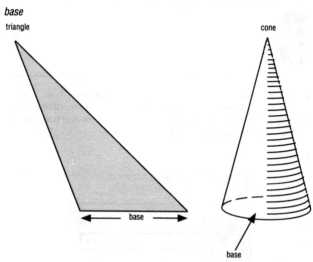

triangle

cone

base

base

A binary number can be converted very simply to an octal (base-eight) number by grouping the digits of the binary number in threes from the right. For example:

binary number 1   110   111   001   becomes

octal number   1   6   7   1

To convert a binary number to a hexadecimal (base-12) number the binary digits are grouped in fours. This economy is utilized in storing data in computer memory.

base in geometry, the line or area on which a polygon or solid stands.

bearing the direction of a fixed point, or the path of a moving object, from a point of observation on the Earth's surface, expressed as an angle from the north. Bearings are taken by ◊compass and are measured in degrees (°), given as three-digit numbers increasing clockwise. For instance, north is 000° and northeast is 045°.

bearing

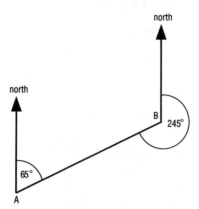

the bearing of B from A is 065°
the backbearing, or bearing of A from B, is 245°

The bearing of one point (B) from another (A) on a map can be found by drawing a line connecting the two points and then drawing another line pointing northwards from A. The angle between the north-pointing line and the line A–B may then be measured with a protractor. The *back bearing* is found by drawing the north line at B and then measuring the clockwise angle from north to the line B–A. There is difference of 180° between a bearing and a back bearing.

best fit see ◊line of best fit.

billion the number represented by a 1 followed by nine zeros (1,000,000,000), equivalent to a thousand million.

Until recently, one billion was taken in the UK to be a million million (1,000,000,000,000), but the first definition is now internationally recognized.

bimodal in statistics, having two distinct peaks of ◊frequency distribution.

binary number system number system to the base two, used in computing and electronics. All binary numbers are written using a combination of the digits 0 and 1. Normal decimal, or base-ten, numbers may be considered to be written under column headings based on the number ten. For example, the decimal number 2,567 stands for:

1,000s	100s	10s	1s
(10^3)	(10^2)	(10^1)	(10^0)
2	5	6	7

Binary, or base-two, numbers may be considered to be written under column headings based on the number two. (Each column has a value twice as great as that of the column to its immediate right.) For example, the binary number 1101 stands for:

8s	4s	2s	1s
(2^3)	(2^2)	(2^1)	(2^0)
1	1	0	1

The binary number 1101 is therefore equivalent to the decimal number 13, since $(1 \times 8) + (1 \times 4) + (1 \times 1) = 13$.

Because binary numbers use only the digits 0 and 1, they can be used as a code to represent instructions or data by any device that can exist in two different states. In a computer several different two-state devices are used to store or transmit binary number codes – for example, circuits, which may or may not carry a voltage; discs or tapes, parts of which may or may not be magnetized; and switches, which may be open or closed.

binomial an algebraic expression that has two ◊variables (denoted by letters); for example, $x + y$ or $x - y$.

Binomial expansions such as

$$(x + y)^2 = x^2 + 2xy + y^2$$

and

$$(x + y)^3 = x^3 + 3x^2y + 3xy^2 + y^3$$

play an important role in statistics and the study of probability. If the probability of an event happening is x and that of the same event not happening is y, then the probabilities for n trials are fully described by the binomial expansion of $(x + y)n$.

Supposing, for example, that when a single die is thrown, a score of 5 or 6 is counted as a winning throw, while a score of 1, 2, 3, or 4 is counted as a losing throw:

probability of a winning throw = probability of scoring 5 + probability of scoring 6 = $^1/_6 + ^1/_6 = ^1/_3$;

probability of a losing throw = probability of scoring 1 + probability of scoring 2 + probability of scoring 3 + probability of scoring 4 = $^1/_6 + ^1/_6 + ^1/_6 + ^1/_6 = ^2/_3$.

For three throws of the die, the probability of each combination of winning and losing throws may be worked out by the expansion of $(^1/_3 + ^2/_3)^3$:

$$(^1/_3 + ^2/_3)^3 = (^1/_3)^3 + 3(^1/_3)^2(^2/_3) + 3(^1/_3)(^2/_3)^2 + (^2/_3)^3$$

probability of three wins	probability of two wins and one loss	probability of one win and two losses	probability of three losses

Therefore, the probability of three wins is 1/27, that of two wins and one loss is 6/27 , that of one win and two losses is 12/27, and that of three losses is 8/27. The sum of these probabilities is 27/27, or 1.

bisect to divide a line or angle into two equal parts.

bisector a line that bisects an angle or another line (known as a *perpendicular bisector* when it bisects at right angles).

block graph in statistics, a diagram in which frequency is represented by height.

bound a number that is larger or smaller than a given set of numbers; see ◊upper bound and ◊lower bound.

boundary another name for ◊perimeter, a line enclosing a shape. When the elements of a ◊set are shown as points within a closed line (for example, in a Venn diagram), the line is called the boundary of that set.

bisect

to bisect a line XY perpendicularly using a pair of compasses

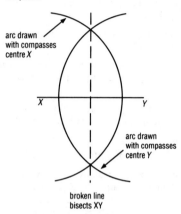

arc drawn
with compasses
centre *X*

arc drawn
with compasses
centre *Y*

broken line
bisects XY

to bisect angle AOB using compasses

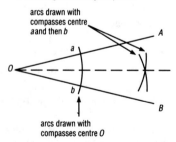

arcs drawn with
compasses centre
a and then *b*

arcs drawn with
compasses centre *O*

broken line
bisects AOB

bracket a sign that shows which part of a calculation should be worked out first. For example $4 \times (7 + 3)$ indicates that 4 is to be multiplied by the result obtained from adding 7 and 3.

breadth thickness, another name for ◊width. The area of a rectangle is given by the formula: area = length × breadth.

C

calculator pocket-sized electronic computing device for performing numerical calculations. It can add, subtract, multiply, and divide; many calculators also compute squares and roots, and have advanced trigonometric and statistical functions. Input is by a small keyboard and results are usually shown on a one-line screen, typically a liquid crystal display (LCD) or a light-emitting diode (LED).

calculus branch of mathematics that permits the manipulation of continuously varying quantities. *Integral calculus* is used to add together the effects of continuously varying quantities. *Differential calculus* deals in a similar way with rates of change. Many of its applications arose from the study of the gradients of the tangents to curves.

cancel to simplify a fraction or ratio by dividing both numerator and denominator by the same number or variable (which must be a ◊common factor of both of them). For example, the algebraic expression $5x/25$ cancels to $x/5$ when divided top and bottom by 5; and $xy^2/2y$ cancels to $xy/2$ when divided top and bottom by y.

capacity the maximum volume of liquid that may be held in a container. Units of capacity include the litre and millilitre (metric), and the pint and gallon (imperial).

cardinal number one of the series of numbers 0, 1, 2, 3, 4, ... Cardinal numbers relate to quantity, whereas ordinal numbers (first, second, third, fourth, ...) relate to order. ◊Integers are used to represent both cardinal and ordinal numbers.

cardioid heart-shaped curve traced out by a point on the circumference of a circle, resulting from the circle rolling around the edge of another circle of the same diameter.

Cartesian coordinate one of a pair of numbers used to define the position of a point by its perpendicular distance from two axes, or

cardioid

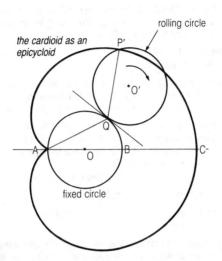

rolling circle

the cardioid as an
epicycloid

P'

O'

Q

A   O   B   C

fixed circle

reference lines, drawn at right angles to each other. For example, a
point P that lies three units from the vertical or *y*-axis and four units
from the horizontal or *x*-axis has the Cartesian coordinates (3, 4).
Cartesian coordinates are named after the 17th-century French mathematician René Descartes.

Celsius scale of temperature, previously called Centigrade, in which
the range from freezing to boiling of water is divided into 100 degrees,
freezing point being zero degrees and boiling point one hundred
degrees.

centimetre metric unit of length (symbol cm) equal to one hundredth
of a metre.

centre of enlargement the focal point of an enlargement, a ⟩trans-
formation that produces a figure of a different size but with similar
dimensions to the original.

chance the likelihood, or ◊probability, of an event taking place, expressed as a fraction or percentage. For example, the chance that a tossed coin will land heads up is 50%.

characteristic the whole-number part of a ◊logarithm. The fractional part is the ◊mantissa.

characteristic feature that distinguishes members of a set from other objects; the rule by which a set is formed. For example, the numbers 10, 20, 50, 80, 100, all have the characteristic that they are multiples of 10.

chord a straight line joining any two points on a curve. The chord that passes through the centre of a circle (its longest chord) is the diameter. The longest and shortest chords of an ellipse (regular oval) are called the major and minor axes respectively.

There are several properties of circles connected with chords:

(1) equal chords subtend equal angles at the centre of a circle;

(2) if two chords of a circle cut each other, the products of the two parts of each chord are equal;

(3) the angle between a tangent and any chord through its point of contact is the same as the angle subtended by the chord at the circumference.

circle a perfectly round shape, the path of a point that moves so as to keep a constant distance from a fixed point (the centre). It is usually drawn with a pair of compasses.

Each circle comprises a *radius* (the distance from any point on the circle to the centre), a *circumference* (the boundary of the circle), *diameters* (lines crossing the circle through the centre), *chords* (lines joining two points on the circumference), *tangents* (lines that touch the circumference at one point only), *sectors* (regions inside the circle between two radii), and *segments* (regions between a chord and the circumference).

The ratio of the distance all around the circle (the circumference) to the diameter is an ◊irrational number called π (*pi*), roughly equal to 3.1416. A circle of radius r and diameter d has a circumference $C = \pi d$, or $C = 2\pi r$, and an area $A = \pi r^2$.

chord

properties of chords

(1) equal chords subtend equal angles at the centre of a circle

AOB and DOC are isosceles triangles (AO, BO, CO, and DO are all radii) and AB = CD, therefore angles AOB and DOC are equal

(2) if two chords cut each other (whether inside or outside a circle), the products of the two parts of each chord are equal

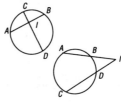

AI × IB = CI × ID

(3) all angles subtended at the circumference by a chord are equal

angles ACB, ADE, AEB, and AFB are equal

circumference the line that encloses a curved plane figure – for example, a ◊circle or an ellipse. Its length varies according to the nature of the curve. The circumference of a circle is given by πd or $2\pi r$, where d is the diameter, r is the radius, and π is the constant pi, approximately equal to 3.1416.

circle

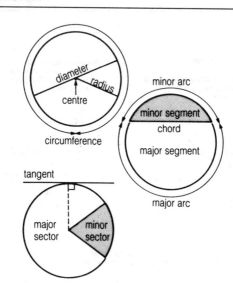

circumscribe to surround a figure with a circle that passes through all the vertices of the figure. Any triangle may be circumscribed and so may any regular polygon. Only certain quadrilaterals may be circumscribed (their opposite angles must add up to 180°).

class another name for a ◊set.

classify to put into separate classes, or ◊sets.

class interval in statistics, the range of each class of data, used when dealing with large amounts of data. To obtain an idea of the distribution, the data are broken down into convenient classes, which must be mutually exclusive and are usually equal. The class interval defines the range of each class; for example if the class interval is five and the data begin at zero, the classes are 0-4, 5-9, 10-14, and so on.

clockwise the direction in which the hands of a clock turn.

closed descriptive of a set of data for which an operation (such as addition or multiplication) done on any members of the set gives a result that is also a member of the set. For example the set of even numbers is closed with respect to addition and multiplication, because the sum and the product of two even numbers are always even numbers. The set of odd numbers on the other hand is closed only for multiplication.

coefficient the number part in front of an algebraic term, signifying multiplication. For example, in the expression

$$4x^2 + 2xy - x$$

the coefficient of x^2 is 4 (because $4x^2$ means $4 \times x^2$), that of xy is 2, and that of x is -1 (because $-1 \times x = -x$).

In general algebraic expressions, coefficients are represented by letters that may stand for numbers; for example, in the equation

$$ax^2 + bx + c = 0$$

a, b, and c are coefficients, which can take any number.

collinear lying on the same straight line.

column a vertical list of numbers or terms, especially in matrices.

combination a selection of a number of objects from some larger number of objects when no account is taken of order within any one arrangement. For example, 123, 213, and 312 are regarded as the same combination of three digits from 1234. Combinatorial analysis is used in the study of ◊probability.

The number of ways of selecting r objects from a group of n is given by the formula

$$\frac{n!}{r!(n-r)!}$$

(where ! is the ◊factorial of a number).

combined probability probability associated with the occurrence of two or more – either of their all taking place at the same time, or of one event taking place but not the others, or of none of the events taking place.

For example, if p_1 is the probability of winning in one lottery and p_2 is the probability of winning in another, the combined probability of winning both lotteries is given by $p_1 \times p_2$ (◊AND rule); the combined probability for winning one lottery or the other but not both is $p_1 + p_2$ (◊OR rule); and the combined probability for not winning either is given by $(1 - p_1)(1 - p_2)$.

common denominator a denominator that is a common multiple of, and therefore exactly divisible by, all the denominators of a set of fractions, and which therefore enables their sums or differences to be found. For example, $^2/_3$ and $^3/_4$ can both be converted to equivalent fractions of denominator 12, $^2/_3$ being equal to $^8/_{12}$ and $^3/_4$ to $^9/_{12}$. Hence their sum is $^{17}/_{12}$ and their difference is $^1/_{12}$. The *lowest common denominator* (lcd) is the smallest common multiple of a given set of fractions.

common difference the difference between any number and the next in an ◊arithmetic progression. For example, in the set 1, 4, 7, 10, ..., the common difference is 3.

common factor a number that will divide two or more others without leaving a remainder. For example, 3 is a common factor of 15, 21, and 24.

common logarithm another name for a ◊logarithm to the base ten.

common multiple a number that is found in the multiplication tables of two or more given numbers. For example, 240 is a common multiple of 6, 8, and 15. However, the ◊lowest common multiple of the three numbers is 120.

commutative operation an operation that is independent of the order of the numbers or symbols concerned. For example, addition is commutative: the result of adding $4 + 2$ is the same as that of adding $2 + 4$; subtraction is not as $4 - 2 = 2$, but $2 - 4 = -2$. Compare ◊associative operation and ◊distributive operation.

compass any instrument for finding direction, or ◊bearing. The most commonly used is a magnetic compass, consisting of a thin piece of magnetic material with the north-seeking pole indicated, free to rotate on a pivot and mounted on a compass card on which the points of the compass are marked. When the compass is properly adjusted and used, the north-seeking pole will point to the magnetic north, from which true north can be found from tables of magnetic corrections.

Also, an instrument used for drawing circles or taking measurements, consisting of a pair of pointed legs connected by a central pivot; otherwise known as a *pair of compasses*.

complement the set of all the elements within the universal set that are not contained in a designated set. For example, if the universal set is the set of all positive whole numbers and the designated set S is the set of all even numbers, then the complement of S (denoted S') is the set of all odd numbers.

complementary angles two angles that add up to 90°.

completing the square method of solving a quadratic equation by converting it into a perfect square – that is, to the form

$$(x + A)^2 = B$$

where A and B are constants. For example, an equation such as

$$x^2 + 6x + 5 = 0$$

is converted into a perfect square by adding the square of half the coefficient of x (that is 3^2, or 9) to each half of the equation.

The steps are:

$$x^2 + 6x = -5$$
$$x^2 + 6x + 9 = -5 + 9$$
$$(x + 3)^2 = 4$$
$$x + 3 = \pm 2$$
$$x = -3 \pm 2$$
$$x = -1 \text{ or } -5$$

Such an equation might also be solved by ◊factorization.

complex number a number written in the form $a + ib$, where a and b are ◊real numbers and i is the square root of –1 (that is, $i^2 = -1$); i used to be known as the 'imaginary' part of the complex number. Some equations in algebra, such as those of the form $x^2 + 5 = 0$, cannot be solved without recourse to complex numbers, because the real numbers do not include square roots of negative numbers.

component one of the the vectors produced when a single vector is resolved into two or more parts. The perpendicular components add up to the original vector.

computer programmable electronic machine that can input, process, store, and output data, and perform calculations and other symbol-manipulation tasks.

computing device any device built to perform or help perform computations, such as the ◊abacus, ◊calculator, or ◊computer.

concave of a surface, curving inwards, or away from the eye. For example, a bowl appears concave when viewed from above. In geometry, a concave polygon is one that has an interior angle greater than 180°. Concave is the opposite of ◊convex.

concave

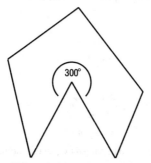

concave polygon

300°

this irregular hexagon is a concave polygon because it
possesses an interior angle that is greater than 180°

concentric circles two or more circles that share the same centre.

conclusion an opinion on, or decision regarding, a problem based on reasoning.

concurrent lines two or more lines passing through a single point; for example, the diameters of a circle are all concurrent at the centre of the circle.

concurrent lines

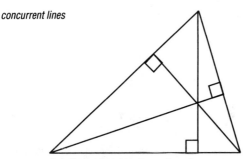

the altitudes of any triangle pass through a common
point and are therefore concurrent

cone a solid figure having a plane (two-dimensional) curve as its base and tapering to a point (the vertex). The line joining the vertex to the centre of the base is called the axis of the cone. A ***circular cone*** has a circle as its base; a cone that has its axis at right angles to the base is called a ***right cone***.

A right circular cone of perpendicular height *h* and base of radius *r* has a volume

$$V = {}^1/_3\pi r^2 h$$

The distance from the edge of the base of a cone to the vertex is called the ***slant height***. In a right circular cone of slant height *l*, the curved surface area is $\pi r l$, and the area of the base is πr^2. Therefore the total surface area is

$$A = \pi r l + \pi r^2 = \pi r(l + r)$$

cone

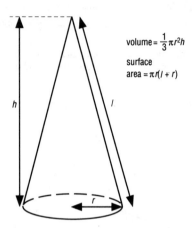

$$\text{volume} = \frac{1}{3}\pi r^2 h$$

surface
area $= \pi r(l + r)$

congruent (of two or more plane or solid figures) having the same shape and size. With plane congruent figures, one figure will fit on top of the other exactly, though this may first require rotation of one of the figures.

The conditions for congruence of a pair of triangles are any of the following:

(1) three sides are equal (SSS);
(2) two sides and an included angle are equal (SAS);
(3) two angles and a corresponding side are equal (AAS);
(4) right-angled, hypotenuse and one other side equal (RHS).

conical having the shape of a cone.

conic section curve obtained when a conical surface is intersected by a plane. Different conic sections are produced depending on where the plane intersects. If the intersecting plane cuts both extensions of the cone, it yields a ◊hyperbola; if it is parallel to the side of the cone, it produces a ◊parabola. Other intersecting planes produce ◊circles or ◊ellipses.

conic section

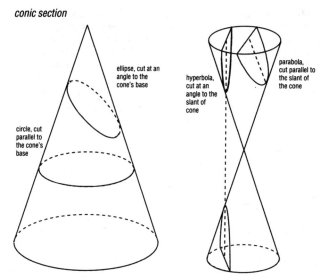

circle, cut parallel to the cone's base

ellipse, cut at an angle to the cone's base

hyperbola, cut at an angle to the slant of cone

parabola, cut parallel to the slant of the cone

conjecture a suggestion based on investigation of some rule or pattern in a problem.

conjugate angles two angles that add up to 360°.

consecutive in logical order. Consecutive numbers are numbers that follow on from each other in order; for example, 3, 4, 5, are consecutive numbers, and so are n, $n + 1$, and $n + 2$.

constant a fixed quantity or one that does not change its value in relation to ♭variables. For example, in the algebraic expression $y^2 = 5x - 3$, the numbers 3 and 5 are constants.

construct in geometry, to draw accurately using drawing instruments.

construction in geometry, a line, angle, or figure that is drawn in order to help solve a problem or produce a proof.

continuous data data that can take an infinite number of values between whole numbers and so cannot be measured completely accurately. This type of data contrasts with ***discrete data***, in which the variable can only take one of a finite set of values. For example, the sizes of apples on a tree form continuous data, whereas numbers of apples form discrete data.

convergence the property of a series of numbers in which the difference between consecutive terms gradually decreases. The sum of a converging series approaches a limit as the number of terms tends to ◊infinity.

converse the reversed order of a conditional statement; the converse of the statement 'if *a*, then *b*' is 'if *b*, then *a*'. The converse does not always hold true; for example, the converse of 'if $x = 3$, then $x^2 = 9$' is 'if $x^2 = 9$, then $x = 3$', which is not true, as *x* could also be –3.

conversion graph graph used for changing values from one unit to another; for example, a temperature reading in the Celsius scale to a reading in the Fahrenheit scale, or German currency (Deutschmarks) to French currency (francs).

conversion graph

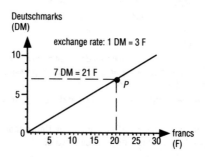

conversion graph for German Deutschmarks and French francs

To construct a graph converting, for example, Deutschmarks to francs, axes are chosen for the two currencies, and the current exchange rate (say, 1 DM = 3 F) used to plot one reference point. The origin (0, 0) is used as a second reference point (because 0 DM = 0 F). A straight-line graph passing through both reference points may then be drawn. A sum in Deutschmarks may now be converted to francs by drawing a horizontal line from, say, the 7 DM mark on the vertical Deutschmark axis to meet the conversion graph at *P*, and then drawing a vertical line down from *P* to the franc axis. The value in francs may then be read from this axis.

conversion table table used for changing values from one unit to another; for example, from imperial to metric measures.

convex of a surface, curving outwards, or towards the eye. For example, the outer surface of a ball appears convex. In geometry, the term is used to describe any polygon possessing no interior angle greater than 180°. Convex is the opposite of ◊concave.

convex

convex polygon

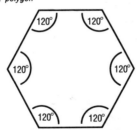

a regular hexagon is a convex polygon: none
of its interior angles is greater than 180°

coordinate in geometry, a number that defines the position of a point relative to a point or axis. ◊Cartesian coordinates define a point by its perpendicular distances from two axes drawn through a fixed point at right angles to each other; ◊polar coordinates define a point in

coordinate

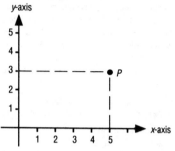

Cartesian coordinates

the Cartesian coordinates of *P* are (5,3)

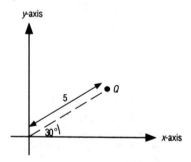

polar coordinates

the polar coordinates of *Q* are (5,30°)

a plane by its distance from a fixed point and direction from a fixed line.

coordinate geometry or *analytical geometry* system of geometry in which points, lines, shapes, and surfaces are represented by algebraic expressions. In plane (two-dimensional) coordinate geometry, the plane is usually defined by two axes at right angles to each other, the horizontal x-axis and the vertical y-axis, meeting at O, the origin. A point on the plane can be represented by a pair of ◊Cartesian coordinates, which define its position in terms of its distance along the x-axis and along the y-axis from O. These distances are respectively the x and y coordinates of the point.

Lines are represented as equations; for example, $y = 2x + 1$ gives a straight line, and $y = 3x^2 + 2x$ gives a ◊parabola (a curve). The graphs of varying equations can be drawn by plotting the coordinates of points that satisfy their equations, and joining up the points.

One of the advantages of coordinate geometry is that geometrical solutions can be obtained without drawing but by manipulating algebraic expressions. For example, the coordinates of the point of intersection of two straight lines can be determined by finding the unique values of x and y that satisfy both of the equations for the lines, that is, by solving them as a pair of ◊simultaneous equations. The curves studied in simple coordinate geometry are the ◊conic sections (circle, ellipse, parabola, and hyperbola), each of which has a characteristic equation.

coplanar in geometry, describing lines or points that all lie in the same plane.

correlation relation between two sets of information: they correlate when they vary together. If one set of data increases at the same time as the other, the relationship is said to be positive or direct. If one set of data increases as the other decreases, the relationship is negative or inverse. Correlation can be shown by plotting a best-fit line on a ◊scatter diagram.

In statistics, such relations are measured by the calculation of ◊coefficients. These generally measure correlation on a scale with 1 indicating perfect positive correlation, 0 no correlation at all, and −1 perfect inverse correlation.

correspondence the relation between two sets where an operation on the members of one set maps some or all of them on to one or more members of the other. For example, if *A* is the set of members of a family and *B* is the set of months in the year, *A* and *B* are in correspondence if the operation is: ' ...has a birthday in the month of... '.

corresponding angles angles that are in matching positions on the same side of a transversal (a line that cuts throught two or more lines in the same plane). Where the lines being cut by the transversal are parallel, the corresponding angles are equal.

cosecant in trigonometry, a ◊function of an angle in a right-angled triangle found by dividing the length of the hypotenuse (the longest side) by the length of the side opposite the angle. Thus the cosecant of an angle *A*, usually shortened to cosec *A*, is always greater than (or equal to) 1. It is the reciprocal of the ◊sine of the angle, that is,

$$\operatorname{cosec} A = \frac{1}{\sin A}$$

cosine in trigonometry, a ◊function of an angle in a right-angled triangle found by dividing the length of the side adjacent to the angle by the length of the hypotenuse (the longest side). Cosine is usually shortened to *cos*.

The two non-right angles of a right-angled triangle add up to 90° and are, therefore, described as *complementary angles* (or co-angles). If the two non-right angles are α and β, it may be seen that

$$\sin \alpha = \cos \beta$$
$$\sin \beta = \cos \alpha.$$

Therefore, the ◊sine of each angle equals the *co*sine of its *co*-angle. For example, if the co-angles of a triangle are 30° and 60°

$$\sin 30° = \cos 60° = 0.5$$
$$\sin 60° = \cos 30° = 0.8660$$

cosine rule a rule of trigonometry that relates the sides and angles of triangles. The rule has the formula

$$a^2 = b^2 + c^2 - 2bc \cos A$$

where *a*, *b*, and *c* are the sides of the triangle, and *A* is the angle opposite *a*.

corresponding angles

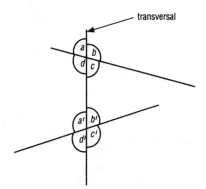

a and *a¹*, *b* and *b¹*, *c* and *c¹*, *d* and *d¹* are pairs of
corresponding angles

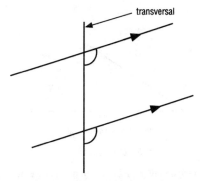

where a transversal cuts through a pair of parallel
lines the corresponding angles are equal

cosine

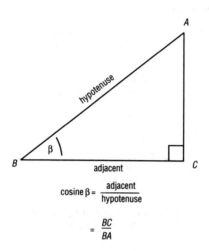

$$\text{cosine } \beta = \frac{\text{adjacent}}{\text{hypotenuse}}$$

$$= \frac{BC}{BA}$$

cosine rule

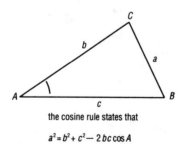

the cosine rule states that

$$a^2 = b^2 + c^2 - 2bc \cos A$$

cotangent in trigonometry, a ◊function of an angle in a right-angled triangle found by dividing the length of the side adjacent to the angle by the length of the side opposite it. It is usually written as *cotan* or *cot*,

contangent

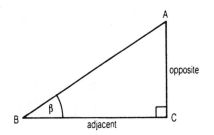

$$\text{cot(angent) } \beta = \frac{1}{\tan \beta} = \frac{\cos \beta}{\sin \beta} = \frac{\text{adjacent}}{\text{opposite}} = \frac{BC}{AC}$$

and it is the reciprocal of the Ďtangent of the angle, so that

$$\cot A = \frac{1}{\tan A}$$

where A is the angle in question.

critical path analysis procedure used in the management of complex projects to minimize the amount of time taken. The analysis shows which subprojects can run in parallel with each other, and which have to be completed before other subprojects can follow on. By identifying the time required for each separate subproject and the relationship between the subprojects, it is possible to produce a planning schedule showing when each subproject should be started and finished in order to complete the whole project most efficiently. Complex projects may involve hundreds of subprojects, and computer programs for critical path analysis are widely used to help reduce the time and effort involved in their analysis.

cross multiply in the case of two fractions, to multiply the numerator of one by the denominator of the other and vice versa. This is usually applied to equivalent fractions, since $a/b = c/d$ if $ad = bc$.

cross section the surface formed when a solid is cut through by a plane at right angles to its axis.

cube in geometry, a solid figure whose faces are all squares. It has six equal-area faces and 12 equal-length edges. If the length of one edge is *l*, the volume *V* of the cube is given by

$$V = l^3$$

and its surface area

$$A = 6l^2$$

cube to multiply a number by itself and then by itself again. For example, 5 cubed = $5^3 = 5 \times 5 \times 5 = 125$. The term also refers to a number formed by cubing; for example, 1, 8, 27, 64 are the first four cubes.

cubic centimetre (or metre) the metric measure of volume, corresponding to the volume of a cube whose edges are all 1 cm (or 1 metre) in length.

cubic equation any equation in which the largest power of *x* is x^3.

cuboid six-sided three-dimensional prism whose faces are all rectangles. A brick is a cuboid.

cumulative frequency in statistics, the total frequency of a given value up to and including a certain point in a set of data. It is used to draw the cumulative frequency curve, the ▷ogive.

cumulative frequency

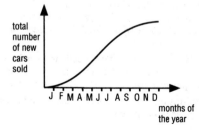

graph showing cumulative sales of new cars throughout the year

total number of new cars sold

J F M A M J J A S O N D

months of the year

curve in geometry, the ◊locus of a point moving according to specified conditions. The circle is the locus of all points equidistant from a given point (the centre). Other common geometrical curves are the ◊ellipse, ◊parabola, and ◊hyperbola, which are also produced when a cone is cut by a plane at different angles.

cusp point where two branches of a curve meet and the tangents to each branch coincide.

cyclic in geometry, describing a polygon of which each vertex (corner) lies on the circumference of a circle. An example is a ◊cyclic quadrilateral. The term is also used in ◊group theory and ◊permutations.

cyclic patterns patterns in which simple ideas are repeated to form more complex designs. Some ◊functions show cyclic patterns, for example mapping round a circle.

cyclic quadrilateral a quadrilateral with all four of its vertices lying on the circumference of a circle. The properties of cyclic quadrilaterals are:

cyclic quadrilateral

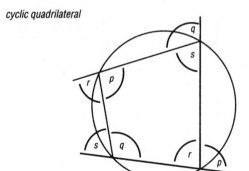

the opposite angles of a cyclic quadrilateral add up to
180° and are therefore said to be supplementary

$$p + r = q + s = 180°$$

(1) the opposite angles add up to 180°, and are therefore said to be sup-plementary;

(2) each external angle (formed by extending a side of the quadrilat-eral) is equal to the opposite interior angle.

cycloid in geometry, a curve resembling a series of arches traced out by a point on the circumference of a circle that rolls along a straight line. Its applications include the study of the motion of wheeled vehi-cles along roads and tracks.

cylinder in geometry, a tubular solid figure with a circular base. In everyday use, the term applies to a *right cylinder*, the curved surface of which is at right angles to the base.

cylinder

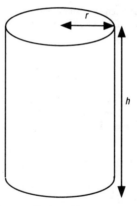

volume = $\pi r^2 h$
area of curved
surface = $2\pi rh$

total surface area
= $2\pi r(r + h)$

The volume V of a cylinder is given by

$$V = \pi r^2 h$$

where r is the radius of the base and h is the height of the cylinder. Its total surface area A has the formula

$$A = 2\pi r(h + r)$$

where $2\pi rh$ is the curved surface area, and $2\pi r^2$ is the area of both circular ends.

cylindrical having the shape of a cylinder. For example, tins of food, toilet rolls, pipes, and poles are all cylindrical objects.

D

data facts, figures, and symbols, especially as stored in computers. The term is often used to mean raw, unprocessed facts, as distinct from information, to which a meaning or interpretation has been applied.

database structured collection of data. The database makes data available to the various computer programs that need it, without the need for those programs to be aware of how the data are stored.

A telephone directory stored as a database might allow all the people whose names start with the letter B to be selected by one program, and all those living in Chicago by another.

decagon in geometry, a ten-sided ◊polygon.

decimal a number less than 1, expressed by figures written after the decimal point. 0.35 and 0.015 are examples of decimals.

decimal fraction a ◊fraction expressed by the use of the decimal point, that is, a fraction in which the denominator is any higher power of 10. Thus $^3/_{10}$, $^{51}/_{100}$, $^{23}/_{1,000}$ are decimal fractions and are normally expressed as 0.3, 0.51, 0.023. The use of decimals greatly simplifies addition and multiplication of fractions, though not all fractions can be expressed exactly as decimal fractions.

decimal number system or *denary number system* the most commonly used number system, to the base ten. Decimal numbers do not necessarily contain a decimal point; 563, 5.63, and –563 are all decimal numbers. Other systems are mainly used in computing and include the ◊binary number system, ◊octal number system, and ◊hexadecimal number system.

Decimal numbers may be thought of as written under column headings based on the number ten. For example, the number 2,567 stands for:

1,000s	*100s*	*10s*	*1s*
(10^3)	(10^2)	(10^1)	(10^0)
2	5	6	7

decimal point the dot dividing a decimal number's whole part from its fractional part. It is usually printed on the line but hand written above the line, for example 3·5. Some European countries use a comma to denote the decimal point, for example 3,56.

deduce to arrive at a conclusion through logic by working from a given statement to its necessary consequence.

definition a statement which describes a mathematical object. If the definition is satisfactory all the properties of the object can be deduced from it.

degree a unit (symbol °) of measurement of an angle or arc. A circle is divided into 360°; a half-turn (the angle on a straight line) is 180°; and a quarter-turn (right angle) is 90°. A degree may be subdivided into 60 minutes (symbol '), and each minute may be subdivided in turn into 60 seconds (symbol ").
  Temperature is also measured in degrees, which are divided on a decimal scale. For example, on the Celsius temperature scale 0° is the freezing point of water, and 100° is its boiling point.

demonstrate to show or explain using examples.

denominator the bottom number of a fraction, so called because it *names* the family of the fraction. The top number, or numerator, specifies how many unit fractions are to be taken.

density measure of the compactness of a substance; it is equal to its mass per unit volume and is measured in kg per cubic metre/lb per cubic foot. Density is a ◊scalar quantity. The density D of a mass m occupying a volume V is given by the formula:

$$D = \frac{m}{V}$$

depth distance from the top downwards or from front to back.

derivative or *differential coefficient* the limit of the gradient of a chord linking two points on a curve as the distance between the points tends to zero; for a function with a single variable, $y = f(x)$, it is denoted by $f'(x)$, $Df(x)$, or dy/dx, and is equal to the gradient of the curve.

determinant an array of elements written as a square, and denoted by two vertical lines enclosing the array. For a 2×2 matrix, the determinant Δ is given by the difference between the products of the diagonal terms. For example, the determinant of

$$\begin{bmatrix} a & b \\ c & d \end{bmatrix} = \begin{vmatrix} a & b \\ c & d \end{vmatrix} = ad - bc$$

Determinants are used to solve sets of ◊simultaneous equations by matrix methods.

   When applied to transformational geometry, the determinant of a 2×2 matrix signifies the ratio of the area of the transformed shape to the original and its sign (plus or minus) denotes whether the image is direct (the same way round) or indirect (a mirror image).

deviation see ◊mean deviation and ◊standard deviation

diagonal a straight line joining one vertex to another (but not one next to it) on a polygon.

   Properties of diagonals include:

(1) the diagonals of a rectangle are equal (useful for checking the straightness of door and window frames);

(2) the diagonals of a rhombus are perpendicular;

(3) the diagonals of a parallelogram bisect each other.

diameter a straight line dividing a circle into two equal halves. Every diameter of a circle passes through the centre.

difference the result obtained when subtracting one number from another. Also, those elements of one ◊set that are not elements of another.

differences the set of numbers obtained from a sequence by subtracting each element from its successor. For example, in the sequence 1, 4,

9, 16, 25 ..., the differences are 3, 5, 7, 9 ... . The patterns of differences are used to analyse sequences.

digit any of the numbers from 0 to 9 in the decimal system. Different bases have different ranges of digits. For example, the ◊hexadecimal system has digits 0 to 9 and A to F, whereas the binary system has two digits (or ◊bits), 0 and 1.

digital root a digit formed by adding the digits of a number. If this leads to a total of 10 or more, the resultant digits are added. For example, finding the digital root of 365 involves the following computations: $3 + 6 + 5 = 14$, followed by $1 + 4 = 5$. The digital root of 365 is therefore 5. Digital roots are valuable in providing independent checks of complicated calculations.

dimension in geometry, the number of measures needed to specify the size of a figure. A point is considered to have zero dimension, a line to have one dimension, a plane figure to have two, and a solid body to have three.

directed number an ◊integer with a positive (+) or negative (–) sign attached; for example, +5 or –5. Directed numbers are commonly seen on thermometer scales and used for temperature readings – for instance, the temperature may be said to have risen from –5°C to +8°C.

On a graph, a positive sign shows a movement to the right or upwards; a negative sign indicates movement downwards or to the left.

discrete data data that take only whole number or fractional values. The opposite is *continuous data*, which can take all in-between values. Examples of discrete data include frequency and population data. However measurements of time and other dimensions can give rise to continuous data.

dispersion in statistics, the extent to which data are spread around a central point (typically the ◊arithmetic mean).

displacement movement from one point to another.

displacement vector a vector which describes how an object has been moved from one position to another.

displacement vector

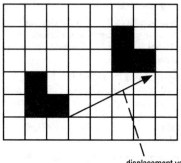

displacement vector

distance–time graph graph used to describe the motion of a body by illustrating the relationship between the distance that it travels and the time taken. Plotting distance (on the vertical axis) against time (on the horizontal axis) produces a graph the gradient, or slope, of which is the body's speed. If the gradient is constant (the graph is a straight line), the body has uniform or constant speed. If the gradient varies (the graph is curved), then so does the speed and the body may be said to be

distance-time graph

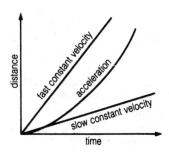

accelerating or decelerating. The speed at any instant is given by the gradient of the tangent at the corresponding point on the curve.

distribution in statistics, the pattern of ◊frequency for a set of data.

distributive operation an operation, such as multiplication, that bears a relationship to another operation, such as addition, such that

$$a \times (b + c) = (a \times b) + (a \times c)$$

For example,

$$3 \times (2 + 4) = (3 \times 2) + (3 \times 4) = 18$$

Multiplication may be said to be distributive over addition. Addition is not, however, distributive over multiplication because

$$3 + (2 \times 4) \neq (3 + 2) \times (3 + 4)$$

The distributive property of numbers is the basis for the long multiplication of two numbers: for example

$$7 \times 25 \rightarrow 7(20 + 5) \rightarrow 140 + 35 \rightarrow 175$$

this is usually laid out as a *sum*:

$$
\begin{array}{r}
25 \\
\times\ 7 \\
\hline
35 \\
140 \\
\hline
175 \\
\end{array}
$$

Compare ◊associative and ◊commutative operation.

dividend any number that is to be divided by another number. For example in the computation $20 \div 4 = 5$, 20 is the dividend.

division basic operation of arithmetic, the inverse of ◊multiplication.

divisor any number that is to be divided into another number. For example in the computation $100 \div 25 = 4$, 25 is the divisor.

dodecahedron a regular solid with 12 pentagonal faces and 12 vertices. It is one of the five regular ◊polyhedra or Platonic solids.

dodecahedron

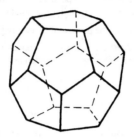

net of a dodecahedron

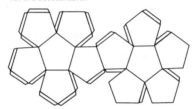

domain the base set of numbers on which a ◊function works, mapping it on to a second set (the ◊range).

duodecimal system system of arithmetic notation using twelve as a base, at one time considered superior to the decimal number system in that 12 has more factors (2, 3, 4, 6) than 10 (2, 5).

E

eccentricity in geometry, a property of a ◊conic section (circle, ellipse, parabola, or hyperbola). It is the distance of any point on the curve from a fixed point (the focus) divided by the distance of that point from a fixed line (the directrix). A circle has an eccentricity of zero; for an ellipse it is less than one; for a parabola it is equal to one; and for a hyperbola it is greater than one.

econometrics the use of mathematics within economics to analyse economic data using statistics.

edge a line along which two planes of a solid figure meet. Also, on a graph, a line joining two nodes or vertices.

element a member of a ◊set.

elevation a drawing to scale of one side of an object or building.

elevation, angle of an upward angle made with the horizontal. For example, the angle of elevation of the Sun starts at zero at dawn, increases to a maximum at noon (zenith), and falls again to zero at

elevation, angle of

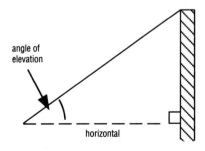

angle of
elevation

horizontal

ellipse

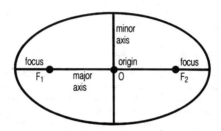

dusk. The angle of elevation at noon may be used by navigators to establish latitude.

ellipse a curve joining all points (loci) around two fixed points (foci) such that the sum of the distances from those points is always constant. The diameter passing through the foci is the major axis, and the diameter bisecting this at right angles is the minor axis. An ellipse is one of a series of curves known as ◊conic sections. A slice across a cone that is not made parallel to, and does not pass through, the base will produce an ellipse.

empty set the set with no elements, symbol ∅ or { }.

enlargement a ◊transformation that produces a figure of a different size, but with similar dimensions to the original. An enlargement may be either smaller or larger.

enlargement

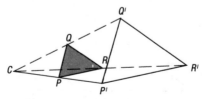

C = centre of enlargement

envelope in geometry, a curve that touches all the members of a family of lines or curves. For example, a family of three equal circles all touching each other and forming a triangular pattern (like a clover leaf) has two envelopes: a small circle that fits in the space in the middle, and a large circle that encompasses all three circles.

epicycloid in geometry, a curve resembling a series of arches traced out by a point on the circumference of a circle that rolls around another circle of a different diameter. If the two circles have the same diameter, the curve is a ◊cardioid.

epicycloid

equal (to) the same (as); of the same size or number.

equation expression that represents the equality of two expressions involving constants and/or variables, and thus usually includes an equals sign (=). For example, the equation

$$A = \pi r^2$$

equates the area A of a circle of radius r to the product πr^2. The algebraic equation

$$y = mx + c$$

is the general one in coordinate geometry for a straight line.

If a mathematical equation is true for all variables in a given domain, it is sometimes called an identity and denoted by $\equiv$. Thus

$$(x + y)^2 \equiv x^2 + 2xy + y^2$$

for all x, y and R.

equilateral of a geometrical figure, having all sides of equal length. For example, a square and a rhombus are both equilateral four-sided figures. An equilateral triangle, to which the term is most often applied, has all three sides equal and all three angles equal (at 60°).

equivalent having a different appearence but the same value. $^3/_5$ and $^6/_{10}$ are equivalent fractions. They both have the value, 0.6.

Eratosthenes' sieve a method for finding ◊prime numbers. It involves writing in sequence all numbers from 2. Then, starting with 2, cross out every second number (but not 2 itself), thus eliminating numbers that can be divided by 2. Next, starting with 3, cross out every third number (but not 3 itself), and continue the process for 5, 7, 11, 13, and so on. The numbers that remain are primes.

error the amount by which an incorrect answer differs from the correct one.

estimate to perform a rough calculation. For example, $17 \div 8 = 2$ roughly.

evaluate to find the value of.

even number any number divisible by 2, hence the digits 0, 2, 4, 6, 8 are even numbers, as is any number ending in these digits, for example 1736. Any whole number which is not even is odd.

event in statistics, any happening to which a probability can be attached.

exceed to be more than. For example, 17 exceeds 12 by 5.

exception a thing that is not subject to a general rule. For example, even numbers cannot be prime numbers (with the exception of 2).

exclusive of events, describing those which cannot happen at the same time. A die, for example, cannot give scores of 2 and 3 in the same throw. The probability of two ◊mutually exclusive events happening together is the product of their separate probabilities.

expand in algebra, to multiply out. For example:

$$ac + bc + ad + bd$$

is the expanded form of

$$(a+b)(c+d)$$

expectation in statistics, the numerical ◊probability of a certain result.

exponent or *index* a number that indicates the number of times a term is multiplied by itself; for example, the exponent of n^2 (or $n \times n$) is two; the exponent of 4^3 (or $4 \times 4 \times 4$) is three.

Exponents obey certain rules. Terms that contain them are multiplied together by adding the exponents; for example:

$$x^2 \times x^5 = x^7$$

The division of such terms is done by subtracting the exponents; for example,

$$y^5 \div y^3 = y^2$$

Any number with the exponent 0 is equal to 1; for example, $x^0 = 1$ and $99^0 = 1$.

exponential descriptive of a ◊function of the form $y = a^x$, that is, in which the variable quantity is an exponent (a number indicating the power to which another number or expression is raised). Population growth and compound interest are both examples of exponential functions.

Exponential functions and series involve the constant e = 2.71828...

expression a mathematical phrase written in symbols. For example, $2x^2 + 3x + 5$ is a ◊quadratic (containing a term or terms raised to the second power but no higher) expression. Equations consist of expressions written around an equals sign.

extend to draw a continuation of a line, usually in order to deduce some property of a figure.

exterior angle one of the four external angles formed when a straight line or transveral cuts through a pair of (usually parallel) lines. Also, an angle formed by extending a side of a polygon.

exterior angle

exterior angles of a transversal

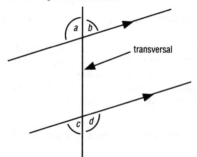

transversal

a, b, c, and *d* are all exterior angles

exterior angles of a polygon

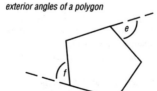

e and *f* are two of the exterior angles of this pentagon

F

face in geometry, a plane surface of a solid enclosed by edges. A cube has six square faces, a cuboid has six rectangular faces, and a tetrahedron has four triangular faces.

factor a number that divides into another number exactly. For example, the factors of 64 are 1, 2, 4, 8, 16, 32, and 64.

factorization the writing of a number or of a ◊polynomial (algebraic expression consisting of more than one variable) as the product of some of its factors. For example, the factors of $x^2 + 3x + 2$ are $x + 1$ and $x + 2$, because

$$x^2 + 3x + 2 = (x + 1)(x + 2)$$

Factorization may be used to solve ◊quadratic equations. For example, to solve the equation

$$x^2 + 6x + 5 = 0$$

the left-hand-side of the equation is factorized as follows,

$$(x + 5)(x + 1) = 0$$

Therefore, either

$$x + 5 = 0 \text{ or } x + 1 = 0$$

$$x = -5 \text{ or } x = -1$$

factorial of a positive number, the product of all the whole numbers (integers) inclusive between 1 and the number itself. A factorial is indicated by the symbol '!'. Thus $6! = 1 \times 2 \times 3 \times 4 \times 5 \times 6 = 720$. Factorial zero, $0!$, is defined as 1.

Fahrenheit scale temperature scale, commonly used in English-speaking countries up until the 1970s, after which the ◊Celsius scale was adopted in line with the rest of the world. In the Fahrenheit scale, intervals are measured in degrees (°F):

$$°F = (°C \times {}^9/_5) + 32$$

fathom in mining and seafaring, a unit of depth measurement (1.83 m/6 ft) used prior to metrication; it approximates to the distance between an adult man's hands when the arms are outstretched.

Fibonacci numbers in their simplest form, a sequence in which each number is the sum of its two predecessors (1, 1, 2, 3, 5, 8, 13, ...). Fibonacci numbers have unusual characteristics with possible applications in botany, psychology, and astronomy.

finite having a countable number of elements, the opposite of infinite.

flow chart diagram, often used in computing, that shows the steps that must be followed to solve a particular problem. The steps are enclosed within boxes and are linked by arrows to show the order in which they must be carried out.

foot imperial unit of length (symbol ft), equivalent to 0.3048 m, in use in Britain since Anglo-Saxon times. It originally represented the length of a human foot. One foot contains 12 inches and is one-third of a yard.

foot the point where a line meets a second line to which it is perpendicular.

formula a set of symbols and numbers which expresses a mathematical factor rule. $A = \pi r^2$ is the formula for calculating the area of a circle. Einstein's famous formula relating energy and mass is $E = mc^2$.

fractal an irregular shape or surface produced by a procedure of repeated subdivision. Generated on a computer screen, fractals are used in creating models for geographical or biological processes (for example, the creation of a coastline by erosion or accretion, or the growth of plants). They are also used in computer art.

fraction a number that indicates one or more equal parts of a whole. Usually, the number of equal parts into which the unit is divided (denominator) is written below a horizontal line, and the number of

flow chart

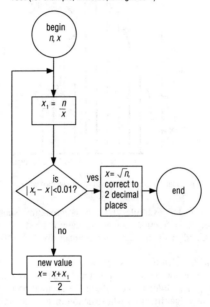

flow chart used to find the square root of any number n, correct to 2 decimal places

the process begins with a rough guess x of the square root (for example, if n is 20, x might be 4)

parts comprising the fraction (numerator) is written above; thus $\frac{2}{3}$ or $\frac{3}{4}$. Such fractions are called *vulgar* or *simple* fractions. The denominator can never be zero.

A *proper fraction* is one in which the numerator is less than the denominator. An *improper fraction* has a numerator that is larger than the denominator – for example, $\frac{3}{2}$. It can therefore be expressed as a

frequency

frequency polygon

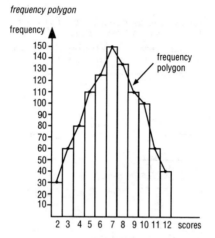

frequency of scores obtained when two dice were
thrown 1, 000 times

mixed number, for example, $1\frac{1}{2}$. A combination such as $\frac{5}{0}$ is
not regarded as a fraction (an object cannot be divided into zero
equal parts), and mathematically any number divided by 0 is equal to
infinity.

A **decimal fraction** has as its denominator a power of 10, and these
are omitted by use of the decimal point and notation, for example 0.04,
which is $\frac{4}{100}$. The digits to the right of the decimal point indicate the
numerators of vulgar fractions whose denominators are 10, 100, 1,000,
and so on. Fractions whose denominators are powers of two or five or
their products can be expressed as exact decimals. In all other cases the
decimal form will have a repeating pattern ($\frac{1}{3} = 0.3333...$ $\frac{43}{99} =$
$0.434343...$ $\frac{1}{7} = 0.14285714...$ etc). The decimal forms of irational
numbers such as $\sqrt{2}$ do not have a repeating pattern, nor do they
terminate.

function

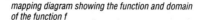

mapping diagram showing the function and domain
of the function f

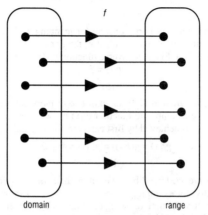

f

domain   range

Fractions are also known as the ***rational numbers***, that is numbers
formed by a ratio. ***Integers*** may be expressed as fractions with a
denominator of 1.

frequency the number of times an event occurs. For example, when
two dice are thrown and the two scores added together, each of the
numbers 2 to 12 has a frequency of occurrence. The set of frequency
data is called a ***frequency distribution***, and is usually presented in a
frequency table or shown diagramatically, as a frequency polygon.

frustum in geometry, a 'slice' taken out of a solid figure by a pair of
parallel planes. A conical frustum, for example, resembles a cone with
the top cut off. The volume and area of a frustum are calculated by subtracting the volume or area of the 'missing' piece from those of the
whole figure.

function rule that maps each element in a given set (the domain) onto
just one element (or image) in another set (the range). For example, the

function $f(x) = x + 3$ maps every number onto one that is larger by three:

$$f(4) = 7$$
$$f(-6) = -3$$

An **inverse function** is responsible for mapping each image in the range back onto the original element in the domain. For example, the function $f(x) = 2x + 1$ has the inverse function $f^{-1}(x) = (x - 1)/2$.

$$f(2) = 5$$
$$f^{-1}(5) = 2$$

A **composite function** is made up of two or more simple functions. For example, the composite function $gf(x)$ of two functions $f(x) = 4x$ and $g(x) = x - 3$ is achieved by first carrying out $f(x)$, and then $g(x)$.

$$gf(x) = g(f(x)) = g(4x) = 4x - 3$$
$$gf(2) = 5$$

Functions are used in all branches of mathematics, physics, and science generally; for example, the formula $t = 2\pi\sqrt{(l/g)}$ shows that for a simple pendulum the time of swing t is a function of its length l and of no other variable quantity (π and g, the acceleration due to gravity, are ◊constants).

G

gallon imperial liquid or dry measure, equal to 4.546 litres, and subdivided into four quarts or eight pints.

generalize to extend a number of results to form a rule. For example, the computations

$$3 + 5 = 5 + 3 \text{ and}$$
$$1.5 + 2.7 = 2.7 + 1.5$$

could be generalized to

$$a + b = b + a$$

generate to produce a sequence of numbers from either the relationship between one number and the next or the relationship between a member of the sequence and its position. For example,

$$u_{n+1} = 2u_n$$

generate

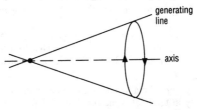

generating a cone by rotating a straight line fixed at one point around a central axis

the cone is the set of positions adopted by the generating line

generates the sequence 1, 2, 4, 8, ... ;

$$an = n(n+1)$$

generates the sequence of numbers 2, 6, 12, 20, ...

In geometry, shapes are generated by moving lines. For example, a cone is generated by a straight line, fixed at one point and rotating around a central axis. A circle is generated by a set of tangents.

geometric mean the nth root of the product of n positive numbers. The geometric mean m of two numbers p and q is such that $m = \sqrt{pq}$. For example, the geometric mean of 2 and 8 is $\sqrt{(2 \times 8)} = \sqrt{16} = 4$.

The geometric mean is always less than the ◊arithmetic mean. This is proved for two numbers as follows:

$$\frac{(a + b)}{2} < \sqrt{ab}$$

therefore,

$$(a + b)^2/_4 < ab$$
$$(a + b)^2 - 4ab < 0$$
$$(a - b)^2 < 0$$

which is always true.

geometric progression or *geometric sequence* a sequence of terms (progression) in which each term is a constant multiple (called the common ratio) of the one preceding it. For example, 3, 12, 48, 192, 768, ... is a geometric sequence with a common ratio 4, since each term is equal to the previous term multiplied by 4. Compare ◊arithmetic progression.

In nature, many single-celled organisms reproduce by splitting in two so that one cell gives rise to 2, then 4, then 8 cells, and so on, forming a geometric sequence 1, 2, 4, 8, 16, 32, ..., in which the common ratio is 2.

geometry branch of mathematics concerned with the properties of space, usually in terms of plane (two-dimensional) and solid (three-dimensional) figures. In ◊coordinate geometry points, curves, and shapes are represented and manipulated by algebraic expressions.

gradient on a graph, the slope of a straight or curved line. The gradient of a curve at any given point is represented by the gradient of the ◊tangent at that point.

Gradient is a measure of rate of change and can be used to represent such quantities as velocity (the gradient of a graph of distance moved against time) and acceleration (the gradient of a graph of a body's velocity against time).

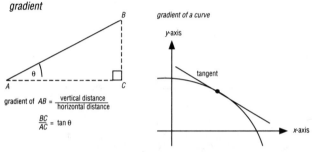

gradient

gradient of AB = $\dfrac{\text{vertical distance}}{\text{horizontal distance}}$

$\dfrac{BC}{AC}$ = $\tan \theta$

gradient of a curve

the gradient to a curve at any point is equal to the gradient of the tangent drawn touching that point

gram metric unit of mass; one-thousandth of a kilogram.

graph pictorial representation of numerical data, such as statistical data, or a method of showing the mathematical relationship between two or more variables by drawing a diagram.

There are often two axes or coordinates at right angles intersecting at the origin – the zero point, from which values of the variables (for example, distance and time for a moving object) are assigned along the axes. Pairs of simultaneous values (the distance moved after a particular time) are plotted as points in the area between the axes, and the points then joined by a smooth curve to produce a graph.

graphical methods methods of solving equations and problems by finding the points of intersection of curves and lines on graphs. For example, the values of x for the equation

graph 68

graph

graph of a straight line

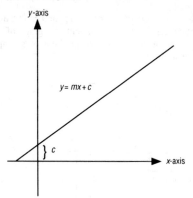

the equation of the straight-line graph takes the form
$y = mx + c$, where m is the gradient (slope) of the line,
and c is the y-intercept (the value of y where the line
cuts the y-axis)

for example, a graph of the equation $y = -x + 4$ will
have a gradient of -1 and will cut the y-axis at $y = 4$

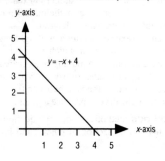

graphical methods

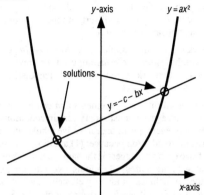

graphical solution of $ax^2 + bx + c = 0$

$$ax^2 + bx + c = 0$$

(which can be rearranged to give $ax^2 = -c - bx$) may be found by finding the points at which the curve

$$y = ax^2$$

cuts the straight-line graph

$$y = -bx - c.$$

The x coordinates of these points are the values of x that satisfy

$$ax^2 = -c - bx.$$

gravity the natural force with which masses are drawn together. The acceleration due to gravity on Earth is 9.8 m/s^2.

great circle circle drawn on a sphere such that the diameter of the circle is a diameter of the sphere. On the Earth, all meridians of longitude are half great circles; among the parallels of latitude, only the equator is a great circle.

The shortest route between two points on the Earth's surface is along the arc of a great circle. These are used extensively as air routes although on maps, owing to the distortion brought about by ◊projection, they do not appear as straight lines.

grid a network of crossing parallel lines. *Rectangular grids* are used for drawing graphs. *Isometric grids* are used for drawing representations of solids in two dimensions in which lengths in the drawing match the lengths of the object.

group a finite or infinite set of elements that can be combined by an operation; formally, a group must satisfy certain conditions. For example, the set of all integers (positive or negative whole numbers) forms a group with regard to addition because: (1) addition is associative, that is, the sum of two or more integers is the same regardless of the order in which the integers are added; (2) adding two integers gives another integer; (3) the set includes an identity element 0, which has no effect on any integer to which it is added (for example, $0 + 3 = 3$); and (4) each integer has an inverse (for instance, 7 has the inverse -7), such that the sum of an integer and its inverse is 0. *Group theory* is the study of the properties of groups.

growth and decay curve graph showing exponential change (growth where the increment itself grows at the same rate) as occurs with compound interest and populations.

H

half one of the parts when a unit is divided into two equal parts; the simplest fraction.

half-turn a transformation in which a figure is rotated through 180°. Reflections and translations can both be achieved by a pair of half-turn transformations.

halve to divide into two equal parts.

hectare metric unit of area equal to 100 ares or 10,000 square metres (2.47 acres), symbol ha.

height of a plane figure or solid, the distance from the top to the base; see ◊altitude.

helix a three-dimensional curve resembling a spring, corkscrew, or screw thread. It is generated by a line that encircles a cylinder or cone at a constant angle.

hemisphere half a sphere, produced when a sphere is sliced along a ◊great circle.

heptagon a seven-sided ◊polygon. It is the shape used at present for 20p and 50p coins.

hexagon a six-sided ◊polygon. The regular hexagon is of great importance in nature as it is the strongest of the plane structures when formed into a lattice.

highest common factor (HCF) in a set of numbers, the highest number which will divide every member of the set without leaving a remainder. For example, 6 is the highest common factor of 36, 48 and 72.

histogram in statistics, a graph showing frequency of data, in which the horizontal axis details discrete units or class boundaries, and the

histogram

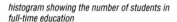

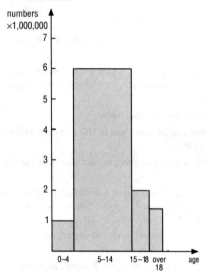

histogram showing the number of students in full-time education

vertical axis represents the frequency. Blocks are drawn such that their areas (rather than their height as in a ◊bar chart) are proportional to the frequencies within a class or across several class boundaries. There are no spaces between blocks.

horizontal having the same direction as the horizon, flat and level. A marble would not roll off a horizontal table without being pushed. On paper, a horizontal line is shown by a line parallel to the top of the page. In practice, horizontal planes are checked with a spirit level.

hour period of time comprising 60 minutes; 24 hours make one calendar day.

hyperbola

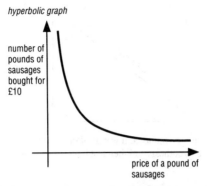

hyperbolic graph

number of
pounds of
sausages
bought for
£10

price of a pound of
sausages

hyperbola in geometry, a curve formed by cutting a right circular cone with a plane so that the angle between the plane and the base is greater than the angle between the base and the side of the cone. All hyperbolae are bounded by two asymptotes (straight lines which the hyperbola moves closer and closer to but never reaches). A hyperbola is a member of the family of curves known as ◊conic sections.

The hyperbola is the curve of indirect proportion; that is, it is the graph formed by two variables that are related in such a way that one increases as the other decreases. Examples of pairs of variables that are indirectly proportional are speed and time of travel over a given distance (the faster the speed, the shorter the time taken), and price and the quantity that can be bought for a fixed sum of money.

hypocycloid in geometry, a cusped curve traced by a point on the circumference of a circle that rolls around the inside of a larger circle.

hypotenuse the longest side of a right-angled triangle, opposite the right angle. It is of particular application in Pythagoras' theorem (the square on the hypotenuse equals the sum of the squares on the other two sides), and in trigonometry where the ratios ◊sine and ◊cosine are defined as the opposite/hypotenuse and adjacent/hypotenuse respectively.

hypotenuse

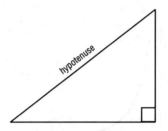

hypothesis an uproven idea. It is part of the process of solving problems: to form a hypothesis and then test it by experiment.

I

icosahedron a twenty-sided ◊polyhedron.

identity a number or operation that leaves others unchanged when combined with them. Zero is the identity for addition; one is the identity for multiplication – for example:

$$7 + 0 = 7, 7 \times 1 = 7$$

The identities are easily shown to be unique: suppose i is the identity for addition and j is another identity, not equal to i. Then

$$i + j = i$$

because j is an identity, but

$$i + j = j$$

because j is an identity. Thus

$$i = j$$

which contradicts what was said above. A similar proof may be carried out for the identity for multiplication.

image a point or number which is produced as the result of a ◊transformation or mapping.

imply to lead logically to. For example, $2x = 10$ implies that $x = 5$. The second statement follows from the first. Implication corresponds to 'if...then' in ordinary language. A single-ended arrow is used for an implication that is not necessarily reversible, and a double-ended arrow for a reversible implication:

$$p \rightarrow q \text{ ... if } p \text{ then } q \text{ (not necessarily reversible);}$$

$$p \leftrightarrow q \text{ ... if } p \text{ then } q \text{ \textbf{and} if } q \text{ then } p.$$

For example:

(1) ABCD is a rectangle → ABCD has equal diagonals;
(2) ABCD is a quadrilateral with all sides equal → ABCD has equal angles (all 90°);
(3) triangle ABC has equal sides ↔ triangle ABC has equal angles;
(4) $x < 5 ↔ x + 2 < 7$.

improper fraction a fraction whose numerator is larger than its denominator.

inch imperial unit of linear measure, a twelfth of a foot, equal to 2.54 centimetres.

include to make one set part of another set. This is possible only if every element in that set belongs to the other set. For example, set A only includes set B if every element of B also belongs to A.

increase to become greater or larger in size or amount. For example, the world population increases by 200 million every year.

independent variable the variable that does not depend on another variable for its values. The symbol x is usually employed to denote the independent variable, while y is used for the dependent variable. Time is always an independent variable.

index in statistics, a numerical scale used to summarise a number of changes and by means of which different levels of data can be compared. For example, the cost of living index number compares the cost of a number of present-day ordinary expenses with the same expenses in previous years.

index (plural *indices*) another term for ◊exponent, the number that indicates the power to which a term should be raised.

inequality a statement that one quantity is larger or smaller than another, employing the symbols > and <. Inequalities may be solved by finding sets of numbers that satisfy them. For example, the solution set to the inequality

$$2x + 5 < 19$$

consists of all values of x less than 7. The symbols ⩾ and ⩽ develop inequality to include the case of equality; for example:

$$x^2 \geqslant 9$$

x must be 3 or greater, or –3 or less.

infinity mathematical quantity that is larger than any fixed assignable quantity; symbol ∞. By convention, the result of dividing any number by zero is regarded as infinity.

inscribed circle a circle drawn inside a plane figure which touches all the sides of the figure. The centre of the inscribed circle of a triangle is the meeting point of the bisectors of the angle of the triangle.

inscribed circle

inscribed circle of a triangle

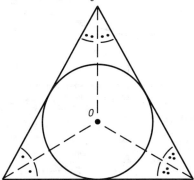

O is the centre of the circle and the meeting point of
the bisectors of each angle in the triangle

integer any whole number. Integers may be positive or negative (0 is an integer and is often considered positive). Fractions, such as $\frac{1}{2}$ and 0.35, are known as ***non-integral numbers*** ('not integers').

intercept the point at which a line or curve cuts across a given axis. The term also refers to the segment cut out of a ◊transversal by the pair of lines that it cuts across.

intercept

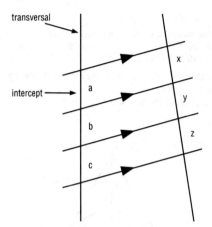

if intercepts a, b, and c are equal, then so are
intercepts x, y, and z

intercepts of a graph

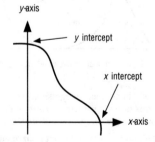

interior angle one of the four internal angles formed when a transversal cuts two or more (usually parallel) lines. Also, one of the angles

interior angle

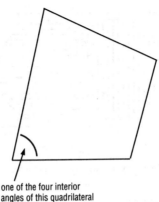

one of the four interior
angles of this quadrilateral

inside a ◊polygon.

interpolation estimate of a value lying between two known values. For example, it is known that 13^2 is 169 and 14^2 is 196. Thus, the square of 13.5 can be interpolated as halfway between 169 and 196.

interquartile range in statistics, a measure of ◊dispersion in a frequency distribution, equalling the difference in value between the upper and lower ◊quartiles.

intersection on a graph, the point where two lines or curves meet. The intersections of graphs provide the graphical solutions of equations.

intersection in set theory, the set of elements that belong to both set A and set B.

interval in statistics, the difference between the smallest and largest measurement in a single set. For example, the interval of normal blood-pressure measurements is 80 – 120 mm of mercury.

inverse the partner of an element that produces the identity when combined with the element. For example, 5 and –5 are additive inverses, and 4 and 0.25 are multiplicative inverses.

intersection of sets

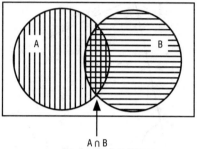

A ∩ B

The inverse of a matrix A is written A^{-1} and is such that

$$AA^{-1} = A^{-1}A = I,$$

where I is the identity matrix.

inverse function a ◊function that exactly reverses the transformation produced by a function f; it is usually written as f^{-1}. Multiplication and division are inverse operations (see ◊reciprocals).

An inverse function is clearly demonstrated on a calculator by entering any number and then pressing x^2 followed by $\sqrt{x}$. The functions on a scientific calculator can be inversed in a similar way.

invert to turn upside down, especially of fractions.

investigation a detailed study of a mathematical situation. For example, a detailed study of the reciprocal numbers 1/1, 1/2, 1/3 ... in decimal form leads to an understanding of the patterns of decimals, and to methods of studying non-repeating decimals such as 0.122333 ... .

involute ◊spiral that can be thought of as being traced by a point at the end of a taut non-elastic thread being wound on to or unwound from a spool.

irrational number a number that cannot be expressed as an exact ◊fraction. Irrational numbers include some square roots (for example,

$\sqrt{2}$, $\sqrt{3}$ and $\sqrt{5}$ are irrational) and numbers such as π (the ratio of the circumference of a circle to its diameter, which is approximately equal to 3.14159).

The proof that $\sqrt{2}$ cannot be a rational number (that is, that it cannot be expressed in the form p/q, where both p and q are whole numbers) is one of the great pieces of mathematics from the ancient world.

To prove $\sqrt{2}$ is irrational, assume $\sqrt{2}$ can be expressed as p/q (p and q have no common factors):

(1) $p/q = \sqrt{2} \to p^2/q^2 = 2 \to p^2 = 2q^2 \to p^2$ is even.

(2) p^2 is even $\to p$ is even $\to p = 2r \to p^2 = 4r^2$

(3) $p^2 = 4r^2 \to 2q^2 = 4r^2 \to q^2 = 2r^2 \to q^2$ is even.

(4) q^2 is even $\to q$ is even.

(5) If both p and q are even, p/q is not in its lowest terms. This contradiction means $p/q = \sqrt{2}$ is false.

isometric paper paper ruled with an isometric ◊grid. It is used for making two-dimensional drawings of three-dimensional objects where lengths are preserved, and is particularly useful in engineering, where objects are constructed from drawings (blueprints).

isometric transformation a transformation in which length is preserved.

isosceles triangle a triangle with two sides equal, hence its base angles are also equal. The triangle has an axis of symmetry which is an ◊altitude of the triangle.

iteration a method of solving equations by a series of approximations which approach the exact solution more and more closely. For example, to find the square root of n, start with a guess x;

(1) calculate $n/x = x_1$;

(2) calculate $(x + x_1)/2 = x_2$;

(3) calculate $n/x_2 = x_3$.

The sequence x_1, x_2, x_3 approaches the exact square root of n. Iterative methods are particularly suitable for work with computers and programmable calculators.

K

kilo- prefix denoting multiplication by 1,000, as in kilohertz, a unit of frequency equal to 1,000 hertz.

kilogram SI unit (symbol kg) of mass equal to 1,000 grams (2.2 lb); also the mass of one litre of water at 4°.

kilometre unit (symbol km) of length equal to 1,000 metres (3,280.89 ft or about 5/8 of a mile).

kite a quadrilateral with two pairs of adjacent equal sides. The geometry of this figure follows from the fact that it has one axis of symmetry.

kite

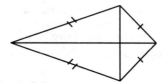

knot unit by which a ship's speed is measured, equivalent to one nautical mile per hour (about 1.85 km/1.15 mi per hour). It is also sometimes used in aviation.

Königsberg bridge problem puzzle that led to the development of topology (the geometry of those properties of a figure which remain the same under distortion). In the city of Königsberg (now Kaliningrad in Russia), seven bridges connect the banks of the River Pregol'a and the islands in the river. For many years, people were challenged to cross each of the bridges in a single tour and return to their starting point. In 1736 Swiss mathematician Leonhard Euler converted the puzzle into a topological network, in which the islands and river banks were represented as nodes (junctions), and the connecting bridges as lines.

Königsberg bridge problem

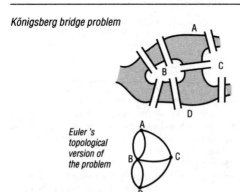

Euler's topological version of the problem

By analysing this network he was able to show that it is impossible to cross each of the bridges once only and return to the point at which one started.

L

latitude and longitude imaginary lines used to locate position on the globe. Lines of latitude are drawn parallel to the equator, with 0° at the equator and 90° at the north and south poles. Lines of longitude are drawn at right angles to these, with 0° (the Prime Meridian) passing through Greenwich, London.

At sea, the latitude is established by the elevation of the Sun or the pole star at its highest position. Longitude is established by the exact time of sunrise and sunset. For centuries longuitude was difficult to establish as chronometers had not been invented and this is why crossings of the oceans were not attempted.

lattice a network of straight lines.

lattice points the points of intersection of the lines in a lattice.

length a measure of the length of an object from end to end. It is the basic dimension of space. The SI unit of length is the metre.

limit in an infinite sequence, the final value towards which the sequence is tending. For example, the limit of the sequence 1/2, 3/4, 7/8, 15/16 ... is 1, although no member of the sequence will ever exactly equal 1 no matter how many terms are added together. The whole subject of calculus is based on the idea of limits.

linear equation an equation involving two variables (x,y) of the general form

$$y = mx + b$$

where m is the slope of the line represented by the equation and b is the y-intercept, or the value of y where the line crosses the y-axis in the ◊Cartesian coordinate system. Sets of linear equations can be used to describe the behaviour of buildings, bridges, trusses, and other static structures.

linear programming in mathematics and economics, a set of techniques for finding the maxima or minima of certain variables governed by linear equations or inequalities. These maxima and minima are used to represent 'best' solutions in terms of goals such as maximizing profit or minimizing cost.

line of best fit on a ◊scatter diagram, line drawn as near as possible to the various points so as to best represent the trend being graphed.

litre metric unit of volume (symbol l), equal to one cubic decimetre (1.76 pints). It was formerly defined as the volume occupied by one kilogram of pure water at 4°C at standard pressure, but this is slightly larger than one cubic decimetre.

locus the path traced out by a moving point. For example, the locus of a point that moves so that it is always at the same distance from another fixed point is a circle; the locus of a point that is always at the same distance from two fixed points is a straight line that perpendicularly bisects the line joining them.

log abbreviation for ◊*logarithm*.

logarithm the ◊exponent or index of a number, usually to the base 10. The logarithm of 1000 is 3 because $10^3 = 1000$; the logarithm of 2 is 0.3010 because $2 = 10^{3,010}$.

loop the part of a curve which encloses a space when the curve crosses itself. In a ◊flow chart, a path which keeps on returning to the same point.

lower bound the value that is less than or equal to all of the values of a given set.

lowest common denominator (lcd) the smallest number that is a multiple of (and therefore exactly divisible by) each of the denominators of a set of fractions. See ◊common multiple.

lowest common multiple (lcm) the smallest number that is a multiple of all the numbers in a given set.

To find the lowest common multiple of a set of numbers, such as 6, 8, and 15, each number must be expressed as the product of its prime factors (factors that can only be divided by 1 or itself). For example,

$$6 = 2 \times 3$$
$$8 = 2 \times 2 \times 2$$
$$15 = 3 \times 5$$

The smallest combination of these prime factors is $2 \times 2 \times 2 \times 3 \times 5 = 120$; therefore the lowest common multiple of 6, 8, and 15 is 120.

M

macro- a prefix meaning on a very large scale, as opposed to ◊micro.

magic square a square array of different numbers in which the rows, columns, and diagonals add up to the same total. A simple example employing the numbers 1 to 9, with a total of 15, is:

$$
\begin{array}{ccc}
6 & 7 & 2 \\
1 & 5 & 9 \\
8 & 3 & 4 \\
\end{array}
$$

magnitude size irrespective of sign, used especially for ◊vectors irrespective of direction.

major arc the larger of the two arcs formed when a circle is divided into two unequal parts by a straight line or chord.

majority the greater number or part of a set. For example, the majority of children like ice cream, the majority of apples are sweet.

mantissa the decimal part of a ◊logarithm. For example, the logarithm of 347.6 is 2.5411; in this case, the 0.5411 is the mantissa, and the integral (whole number) part of the logarithm, the 2, is the ◊characteristic.

mapping or *map* a rule that links elements from one set with those of another; see ◊function and ◊transformation, specific types of mapping.

mass the amount of matter in a body. Mass determines the acceleration produced in a body by a given force acting on it, the acceleration being inversely proportional to the mass of the body. It also determines the force exerted on a body by gravity on Earth (the body's weight). It is the force of gravity acting upon mass that

mapping

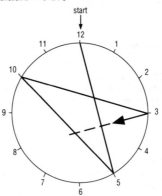

mapping around the face of a clock with the function n ⟶ n + 5

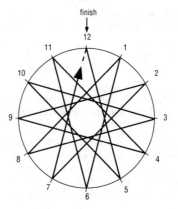

produces weight. Thus, the weight of an object on the Moon will be less than on Earth, while its mass remains the same.

In the SI system, the base unit of mass is the kilogram. Mass is a ◊scalar quantity.

mathematical induction formal method of proof in which the proposition $P(n + 1)$ is proved true on the hypothesis that the proposition $P(n)$ is true. The proposition is then shown to be true for a particular value of n, say k, and therefore by induction the proposition must be true for $n = k + 1, k + 2, k + 3, ...$. In many cases $k = 1$, so then the proposition is true for all positive integers.

mathematics science of spatial and numerical relationships. The main divisions of *pure mathematics* include geometry, arithmetic, algebra, calculus, and trigonometry. Mechanics, statistics, numerical analysis, computing, the mathematical theories of astronomy, electricity, optics, thermodynamics, and atomic studies come under the heading of *applied mathematics*.

matrix a square ($n \times n$) or rectangular ($m \times n$) array of elements (numbers or algebraic variables). Matrices are a means of condensing information about mathematical systems and can be used for, among other things, solving ◊simultaneous linear equations and transformations. Matrices with only one column are called *column vectors*, and those with only one row are called *row vectors*.

Two matrices are multiplied by combining the rows and columns of each matrix (this is only possible if the second matrix has the same number of columns as the first has rows). A new matrix is formed in which the elements are products of the rows of the first matrix and the columns of the second. For example, if

$$A = \begin{bmatrix} a & b \\ c & d \end{bmatrix}$$

and

$$B = \begin{bmatrix} p & q \\ r & s \end{bmatrix}$$

then

$$AB = \begin{bmatrix} ap + br & aq + bs \\ cp + dr & cq + ds \end{bmatrix}$$

Matrix multiplication is associative; that is,

$$(AB)C = A(BC)$$

However, it is not commutative,

$$AB \neq BA$$

identity matrix An identity matrix I is a matrix that leaves other matrices unchanged when it is combined with them. For example, I for the multiplication of a 2×2 matrix is

$$\begin{bmatrix} 1 & 0 \\ 0 & 1 \end{bmatrix}$$

inverse matrix Each matrix has an inverse A^{-1} such that

$$AA^{-1} = A^{-1}A = I$$

provided that $|A|$ (the ◊determinant of A) is not zero.

maximum and minimum in ◊coordinate geometry, points at which the slope of a curve representing a ◊function changes from positive to negative (maximum), or from negative to positive (minimum). A tangent to the curve at a maximum or minimum has zero gradient.

mean a measure of the average of a number of terms or quantities. The simple *arithmetic mean* is the average value of the quantities, that is, the sum of the quantities divided by their number. The *weighted mean* takes into account the frequency of the terms that are summed; it is calculated by multiplying each term by the number of times it occurs, summing the results and dividing this total by the total number of occurrences. The *geometric mean* of n quantities is the nth root of their product. In statistics, it is a measure of central tendency of a set of data.

mean deviation in statistics, a measure of the spread of a population from the ◊mean.

measure to find out size by comparing with a standard.

maximum and minimum

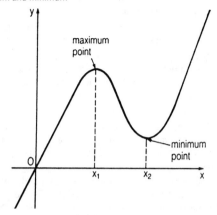

median the middle number of an ordered group of numbers. If there is no middle number (because there is an even number of terms), the median is the ◊mean (average) of the two middle numbers. For example, the median of the group 2, 3, 7, 11, 12 is 7; that of 3, 4, 7, 9, 11, 13 is 8 (the average of 7 and 9).

median in geometry, a line from the vertex of a triangle to the midpoint of the opposite side.

median

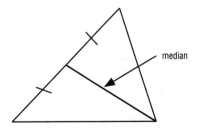

mega- prefix denoting multiplication by a million. For example, a megawatt (MW) is equivalent to a million watts.

member one of the elements belonging to a set. For example, 25 and 2,500 are both members of the set of square numbers, but 250 is not.

memory in a ◊calculator, the function which saves numbers for future use. Calculators usually have two memory keys. One stores the number on display while the other brings the number in the memory back to the display.

mensuration the science of measurement.

metre SI unit (symbol m) of length, equivalent to 1.093 yards. It is defined by scientists as the length of the path travelled by light in a vacuum during a time interval of 1/299,792,458 of a second.

metric system system of weights and measures developed in France in the 18th century and recognized by other countries in the 19th century. In 1960 an international conference on weights and measures recommended the universal adoption of a revised SI system (Système International d'Unités), with seven prescribed 'base units', including the metre (m) for length, the kilogram (kg) for mass, and the second (s) for time.

Prefixes used with metric units are *tera* (T), million million times; *giga* (G), billion (thousand million) times; *mega* (M), million times; *kilo* (k), thousand times; *hecto* (h), hundred times; *deka* (da), ten times; *deci* (d), tenth part; *centi* (c), hundredth part; *milli* (m), thousandth part; *micro* (μ), millionth part; *nano* (n), billionth part; *pico* (p), trillionth part; *femto* (f), quadrillionth part; *atto* (a), quintillionth part.

micro- prefix (symbol μ) denoting a one-millionth part (10^{-6}). For example, a micrometre, m, is one-millionth of a metre.

mile imperial unit of linear measure. A statute mile is equal to 1,760 yards (1.60934 km), and an international nautical mile is equal to 2,026 yards (1,852 m).

milli- prefix (symbol m) denoting a one-thousandth part (10^{-3}). For example, a millimetre, mm, is one thousandth of a metre.

million one thousand thousands, 1,000,000 or 10^6.

minimum see ◊maximum and minimum.

minor arc the smaller of the two arcs formed when a circle is divided into two unequal parts by a straight line or chord. See ◊arc.

minus sign the sign (symbol –) indicating subtraction or denoting a negative number.

minute basic unit of time. There are 60 minutes in one hour and 60 seconds in one minute.

minute or *arc minute* of an angle, a unit equal to one-sixtieth of a degree, symbol '. A minute may be further divided into 60 seconds.

mixed number a number consisting of both whole and fractional parts. For example, $3\frac{1}{2}$ and $7\frac{3}{8}$ are mixed numbers. When calculating with mixed numbers, it is usual to deal with the whole number parts first, or else to convert them all to decimals or improper fractions.

Möbius strip structure made by giving a half twist to a flat strip of paper and joining the ends together. It has certain remarkable properties, arising from the fact that it has only one edge and one side. If cut down the centre of the strip, instead of two new strips of paper, only one long strip is produced.

   A Möbius strip is often used in belt-driven machines: the twist in the belt ensures that both sides of the belt are used, and reduces wear by 50%.

mode the element that appears most frequently in a given group. For example, the mode of the group 0, 0, 9, 9, 9, 12, 87, 87 is 9. (Not all groups have modes.)

   As a statistic, the mode is used commercially to decide which sector of the community should be the target of advertising. For example, if it were found that the modal age range of people buying a washing machine was 25–30, the advertisements for washing machines would all feature people of that age.

modulus another name for ◊absolute value.

multiple any number that is the product of a given number. If *N* appears in the multiplication table of *n*, it is a multiple of *n*; it follows from this that *n* is a factor of *N*. For example, 18 is a multiple of 3 since it appears in the multiplication table of 3, and 3 is a factor of 18 for the same reason. The *lowest common multiple* (lcm) of a set of numbers is the lowest number which is a multiple of them all.

multiplication one of the four basic operations of arithmetic, usually written in the form $a \times b$ or ab, and involving repeated addition in the sense that a is added to itself b times. Multiplication obeys ◊commutative,◊associative, and ◊distributive laws and every number (except 0) has a multiplicative inverse. The number 1 is the ◊identity for multiplication.

multiplication table the set of multiples of a given number obtained by multiplying it by 1, 2, 3 etc. The tables for the first ten numbers are usually grouped into a square, known as the multiplication table.

mutually exclusive describing two events that cannot happen together. It is not possible to throw heads and tails with the same coin at the same time. The combined probability that either one or the other of two mutually exclusive events will happen is the sum of their separate probabilities (◊OR rule). Obviously the probability that both events will happen is zero. For example, if I buy 5 tickets out of 1,000 in a lottery with only one winning ticket, the chance of winning the prize is $0.01 + 0.01 + 0.01 + 0.01 + 0.01 = 0.05$

N

natural number one of the set of numbers used for counting. Natural numbers comprise all the ⏵positive integers, excluding zero.

negative integer or *negative number* any ⏵real number with a value of less than zero. They obey all the usual rules for ⏵positive numbers, resulting in the product of two negative numbers being positive. Every positive number has its negative partner, their sum being zero.

net in geometry, a plan which can be used to make a model of a solid.

net

the net of a cube

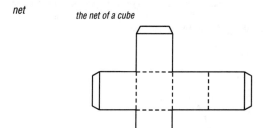

network system of ⏵nodes (junctions) and ⏵arcs (transport routes) through which goods, services, people, money, or information flow. Networks are often shown on topological maps.

node point where routes meet. It may therefore be the same as a ⏵route centre. In a topological ⏵network, a node may be the start or crossing point of routes, also called a *vertex*.

normal distribution curve the distinctive bell-shaped curve obtained when continuous variation within a population is expressed

network

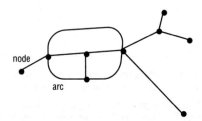

graphically. When a statistician studies height or intelligence, most people have an intermediate or 'normal' score, with a few individuals scoring either high or low. It is by no means certain that intelligence is really fitted to a normal distribution curve as people are often more intelligent in one area of activity than in another; the normal distribution gives a model that may not be a true picture.

normal distribution curve

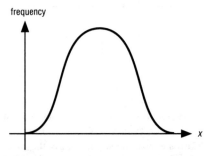

number symbol used in counting or measuring. There are various kinds of numbers. The everyday number system is the decimal ('proceeding by tens') system, using the base ten. ◊Real numbers include all rational numbers (integers, or whole numbers, and fractions) and irrational numbers (those not expressible as fractions). ◊Complex numbers are numbers of the form $a + ib$ where a, b are real numbers and i is $\sqrt{-1}$. Complex numbers have real numbers as a subset (for which $b = 0$). The ◊binary number system, used in computers, has two as its base.

The ordinary numerals, 0, 1, 2, 3, 4, 5, 6, 7, 8, and 9, give a counting system that, in the decimal system, continues 10, 11, 12, 13, and so on. These are whole numbers (positive integers), with fractions represented as, for example, $^1/_4$, $^1/_2$, $^3/_4$, or as decimal fractions (0.25, 0.5, 0.75). They are also rational numbers. Irrational numbers cannot be represented in this way and require symbols, such as $\sqrt{2}$, π, and e. They can be expressed numerically only as the (inexact) approximations 1.414, 3.142 and 2.718 (to three places of decimals) respectively.

numerator the top number of a fraction.

O

observation sheet in statistics, specially prepared record sheet for experiments, including space for tally marks and for total frequency.

obtuse angle an angle greater than 90° but less than 180°.

octagon a ◊polygon with eight sides.

octahedron, regular a regular solid comprised of eight faces, each of which is an equilateral triangle. It is one of the five regular polyhedra or Platonic solids. The figure made by joining the midpoints of the faces is a perfect cube and the vertices of the octahedron are themselves the midpoints of the faces of a surrounding cube. For this reason, the cube and the octahedron are called dual solids.

octahedron

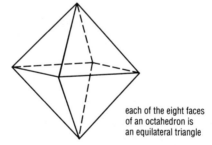

each of the eight faces
of an octahedron is
an equilateral triangle

odd describing any number not divisible by 2, thus odd numbers form the infinite sequence 1, 3, 5, 7 ... . Every square number n^2 is the sum of the first n odd numbers. For example:

$$49 = 7^2 = 1 + 3 + 5 + 7 + 9 + 11 + 13$$

ogive in statistics, the curve on a graph representing ◊cumulative frequency.

operation action on numbers, matrices, or vectors that combines them to form others. The basic operations on numbers are addition, subtraction, multiplication, and division.

Matrices involve the same operations as numbers, but two different types of multiplication are used for vectors: vector multiplication and scalar multiplication. In the scalar multiplication of vectors, the product is a pure number and is the sum of the products of the components. In vector multiplication, the product is a new vector, at right angles to the plane defined by the other two.

opposite side in a right-angled triangle, the side opposite a given angle, but not the hypotenuse (the longest side opposite the right angle). The ◊sine ratio is equal to opposite/hypotenuse.

orbit the path that one heavenly body follows around another, or the path of a satellite travelling around the Earth.

order to arrange with regard to size or quantity or other quality – for example, alphabetical order. Putting a set of things in order is the same as mapping them onto the set of natural numbers. Order relations are shown using the symbols $>$, $<$, $\leq$, and $\geq$. If numbers are being ordered $a > b$ means a is to the left of b on the left-to-right number line. Negative numbers can also be ordered on the number line, bearing in mind that $-a > -b \rightarrow a < b$. For example, -20 is *less than* -10 even though 20 is a larger number than 10.

ordered pair a pair of numbers whose order makes a difference to their meaning. Coordinates are an ordered pair because the point (2,3) is not the same as the point (3,2). Vulgar fractions are ordered pairs because the top number gives the quantity of parts while the bottom gives the number of parts into which the unit has been divided.

ordinal number one of the series first, second, third, fourth, ... Ordinal numbers relate to order, whereas ◊cardinal numbers (1, 2, 3, 4, ...) relate to quantity, or count.

ordinate in ◊coordinate geometry, the y-coordinate of a point; that is, the vertical distance of the point from the horizontal or x-axis. For example, a point with the coordinates (3,4) has an ordinate of 4.

The x-coordinate of a point is known as the ◊abscissa.

ordinate

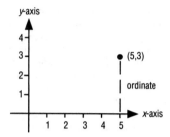

origin the point where the *x* axis meets the *y* axis. The coordinates of the origin are (0,0).

OR rule rule used for the finding the combined probability of one event or another taking place. If two events E_1 and E_2 are mutually exclusive (cannot happen at the same time) and the probabilities of their taking place are p_1 and p_2, respectively, then the probability *p* that either E_1 *or* E_2 will happen is given by

$$p = p_1 + p_2$$

For example, if a blue die and a red die are thrown together, the probability of a blue six is $^1/_6$, and the probability of a red six is $^1/_6$. Therefore, the probability of either a red six or a blue six being thrown is $^1/_6 + ^1/_6 = ^2/_6 = ^1/_3$.

By contrast, the ◊AND rule is used for finding the probability of two or more events taking place together.

ounce imperial measure of weight (symbol oz), equivalent to 28.35 grams. There are sixteen ounces in one pound.

outcome in ◊probability theory or statistics, one possible result of an experiment. For example, the possible outcomes from a match between team A and team B are (i) A wins (ii) B wins and (iii) a draw.

oval egg shaped; see ◊ellipse.

P

parabola a curve formed by cutting a right circular cone with a plane parallel to the sloping side of the cone. A parabola is one of the family of curves known as ◊conic sections. The graph of $y = x^2$ is a parabola.

It can also be defined as a path traced out by a point that moves in such a way that the distance from a fixed point (focus) is equal to its distance from a fixed straight line (directrix).

Parabolic mirrors are used in telescopes and searchlights because parallel rays are reflected to the focus of the parabola.

parabola

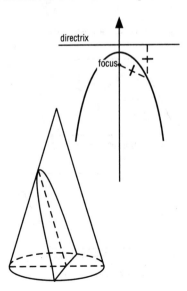

parallel lines and parallel planes straight lines or planes that always remain a constant distance from one another no matter how far they are extended.

parallelogram a quadrilateral (four-sided plane figure) with opposite pairs of sides equal in length and parallel, and opposite angles equal. The diagonals of a parallelogram bisect each other. Its area is the product of the length of one side and the perpendicular distance between

parallelogram

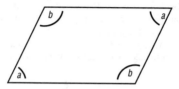

(i) opposite sides and anngles are equal

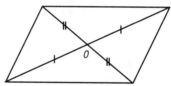

(ii) diagonals bisect each other at *O*

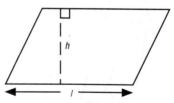

(iii) area of a parallelogram *l* × *h*

this and the opposite side. In the special case when all four sides are equal in length, the parallelogram is known as a rhombus, and when the internal angles are right angles, it is a rectangle or square.

parameter variable factor or characteristic. For example, length is one parameter of a rectangle; its height is another. In computing, it is frequently useful to describe a program or object with a set of variable parameters rather than fixed values. For example, if a programmer writes a routine for drawing a rectangle using general parameters for the length, height, line thickness, and so on, any rectangle can be drawn by this routine by giving different values to the parameters.

partition the division of a set into parts.

Pascal's triangle a triangular array of numbers (with 1 at the apex), in which each number is the sum of the pair of numbers above it. The horizontal rows of Pascal's triangle have totals that are powers of 2, for example $1 + 6 + 15 + 20 + 15 + 6 + 1 = 64 = 2^6$. When plotted at equal distances along a horizontal axis, the numbers in the rows give the ◊binomial probability distribution (with equal probability of success and failure) of an event, such as tossing coins.

Pascal's triangle

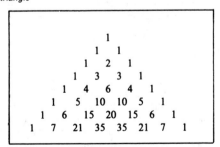

pattern a regular design comprising shapes or numbers.

pentagon a five-sided plane figure. The five-pointed star formed by drawing all the diagonals of a regular pentagon is called a *pentagram*.

percentage way of representing a number as a ◊fraction of 100. Thus 45 percent (45%) equals 45/100, and 45% of 20 is

$$\frac{45}{100} \times 20 = 9$$

In general, if a quantity x changes to y, the percentage change is

$$\frac{100(y - x)}{x}$$

Thus, if the number of people in a room changes from 40 to 50, the percentage increase is

$$\frac{(100 \times 10)}{40} = 25\%$$

To express a fraction as a percentage, it must first be converted to a fraction with a denominator of 100, for example,

$$\frac{1}{8} = \frac{12.5}{100} = 12.5\%$$

The use of percentages often makes it easier to compare fractions that do not have a common denominator.

To convert a fraction to a percentage on your calculator, divide numerator by denominator. The percentage will correspond to the first figures of the decimal, for example

$$\frac{7}{12} = 0.5833333 = 58.3\%$$

correct to three decimal places.

percentile in a frequency distribution, one of the 99 values of a variable that divide its distribution into 100 parts of equal frequency. In practice, only certain of the percentiles are used. They are the ◊median (or 50th percentile), the lower and the upper ◊quartiles (respectively, the 25th and 75th percentiles), the 10th percentile which cuts off the bottom 10% of a frequency distribution, and the 90th which cuts off the top 10%. The 5th and 95th are also sometimes used.

perimeter or *boundary* a line drawn around the edge of an area or shape; also the length of that line. For example, the perimeter of a

perimeter

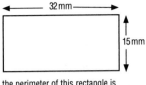

the perimeter of this rectangle is
(32 + 15 + 32 + 15)mm = 94mm

rectangle is the sum of its four sides; the perimeter of a circle is known
as its ***circumference***.

permutation a specified arrangement of a group of objects. It is the
arrangement of *a* distinct objects taken *b* at a time in all possible orders.
It is given by

$$\frac{a!}{(a-b)!}$$

where '!' stands for ◊factorial. For example, the number of per-
mutations of four letters taken from any group of six different letters
is

$$\frac{6!}{2!} = \frac{(1 \times 2 \times 3 \times 4 \times 5 \times 6)}{(1 \times 2)} = 360$$

The theoretical number of four-letter 'words' that can be made from an
alphabet of 26 letters is 26!/22! = 358,800. An alternative way of look-
ing at the same problem is to consider the possible number of choices
for each letter of the four-letter word. There are 26 choices for the first
letter, leaving only 25 for the second, and so on.

See also ◊combination.

perpendicular at a right angle; also, a line at right angles to another
or to a plane.

For a pair of *skew* lines (lines in three dimensions that do not meet),
there is just one common perpendicular, which is at right angles to both
lines; the nearest points on the two lines are the feet of this perpendicular.

perpendicular

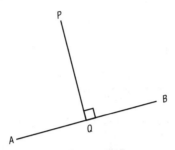

PQ is the perpendicular to *AB*

perpendicular bisector a straight line perpendicular to a line segment and passing through its mid-point. The locus of points equidistant from two points *P* and *Q* is the perpendicular bisector and the axis of symmetry of the line segment *PQ*.

perspective the realistic representation of a three-dimensional object in two dimensions. In a perspective drawing, vertical lines are drawn parallel from the top of the page to the bottom. Horizontal lines, however, are represented by straight lines which meet at one of two perspective points. These perspective points lie to the right and left of the drawing at a distance which depends on the view being taken of the object.

perspective

perspective drawing of a cuboid

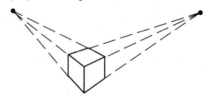

pi symbol π, the ratio of the circumference of a circle to its diameter. The value of pi is 3.1415926, correct to seven decimal places. Common approximations to pi are $^{22}/_7$ and 3.14.

pictogram a pictorial way of presenting statistical data, in which a symbol is used to represent a specific quantity of items.

pie chart method of displaying proportional information by dividing a circle up into different-sized sectors (slices of pie). The angle of each sector is proportional to the size, expressed as a percentage, of the group of data that it represents.

For example, data from a traffic survey could be presented in a pie chart in the following way:

(1) convert each item of data to a percentage figure;
(2) 100% will equal 360 degrees of the circle, therefore each 1% = 360/100 = 3.6 degrees;
(3) calculate the angle of the segment for each item of data, and plot this on the circle.

The diagram may be made clearer by adding colours or shadings to each group, together with a key.

pie chart

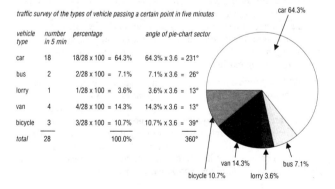

traffic survey of the types of vehicle passing a certain point in five minutes

vehicle type	number in 5 min	percentage		angle of pie-chart sector	
car	18	18/28 x 100 =	64.3%	64.3% x 3.6 =	231°
bus	2	2/28 x 100 =	7.1%	7.1% x 3.6 =	26°
lorry	1	1/28 x 100 =	3.6%	3.6% x 3.6 =	13°
van	4	4/28 x 100 =	14.3%	14.3% x 3.6 =	13°
bicycle	3	3/28 x 100 =	10.7%	10.7% x 3.6 =	39°
total	28		100.0%		360°

car 64.3%

van 14.3%   bus 7.1%

bicycle 10.7%   lorry 3.6%

place value the value given to a digit because of its position within a number. For example, in the decimal number 2,465 the 2 represents two thousands, the 4 represents four hundreds, the 6 represents six tens, and the 5 represents five units.

plan a scale drawing of an object viewed from above.

plane a flat surface. Planes are either parallel or they intersect in a straight line. Vertical planes, for example the join between two walls, intersect in a vertical line. Horizontal planes do not intersect since they are all parallel.

plane figure in geometry, a two-dimensional figure, with height and width but no depth. All ◊polygons are plane figures.

plot to mark the points corresponding to the number pairs, or Cartesian coordinates, of a function onto a graph.

point in geometry, a basic element, whose position in the Cartesian system may be determined by its ◊coordinates.

Mathematicians have had great difficulty in defining the point, as it has no size, and is only the place where two lines meet. According to the Greek mathematician Euclid, (i) a point is that which has no part; (ii) the straight line is the shortest distance between two points.

polar coordinates a way of defining the position of a point in terms of its distance r from a fixed point (the origin) and its angle θ degrees or radians to a fixed line or axis. The coordinates of the point are (r,θ).

The polar-coordinate system is useful for defining positions in programming the operations of, for example, computer-controlled cloth- and metal-cutting machines.

polygon in geometry, a plane (two-dimensional) figure with three or more straight-line sides. Common polygons have names which define the number of sides (for example, triangle, quadrilateral, pentagon).

These are all convex polygons, having no interior angle greater than 180°. The sum of the internal angles of a polygon having n sides is given by the formula $(2n-4) \times 90°$; therefore, the more sides a polygon has, the larger the sum of its internal angles and, in the case of a convex polygon, the more closely it approximates to a circle.

polyhedron in geometry, a solid figure with four or more plane faces. The more faces there are on a polyhedron, the more closely it approximates to a sphere. Knowledge of the properties of polyhedra is needed in crystallography and stereochemistry to determine the shapes of crystals and molecules.

There are only five types of regular polyhedron (with all faces the same size and shape); they are the tetrahedron (four equilateral triangular faces), cube (six square faces), octahedron (eight equilateral triangles), dodecahedron (12 regular pentagons) and icosahedron (20 equilateral triangles).

polynomial an algebraic expression that has one or more ◊variables (denoted by letters). A polynomial of degree one, that is, whose highest ◊power of x is 1, as in $2x + 1$, is called a linear polynomial; $3x^2 + 2x + 1$ is quadratic; $4x^3 + 3x^2 + 2x + 1$ is cubic.

population in statistics, the ◊universal set from which a sample of data is selected. The chief object of statistics is to find out population characteristics by taking ◊samples.

position vector a vector that defines the position of a point.

positive denoting greater than zero. Positive directions are, by convention, from left to right and from the bottom upwards.

positive integer any whole number from 0 upwards. 0 is included so that the properties of positive integers include those related to the ◊identity for addition.

pound imperial unit of weight (symbol lb), equivalent to 454 grams (1 kg = 2.205 lb). A pound may be divided into sixteen ounces. Other imperial units are related as follows: 14 pounds = 1 stone (used in personal bodyweight); 112 pounds = 1 hundredweight (cwt); 2,240 pounds = 1 ton.

power that which is represented by an ◊exponent or index, denoted by a superior small numeral. A number or symbol raised to the power of 2, that is, multiplied by itself, is said to be squared (for example, 3^2, x^2), and when raised to the power of 3, it is said to be cubed (for example, 2^3, y^3).

prime factor any factor of a number which is a prime number. The fundamental theorem of arithmetic states that every number is either prime or can be expressed as a unique product of primes.

prime number a number that can be divided only by 1 or itself, that is, having no other factors. There is an infinite number of primes, the first ten of which are 2, 3, 5, 7, 11, 13, 17, 19, 23, and 29 (by definition, the number 1 is excluded from the set of prime numbers). The number 2 is the only even prime because all other even numbers have 2 as a factor.

Over the centuries mathematicians have sought general methods (algorithms) for calculating primes, from ◊Eratosthenes' sieve to programs on powerful computers. The Greek mathematician Euclid proved that there cannot be a largest prime number. The proof is given below:

(1) Let *M* be the largest prime.
(2) Form a new number *N* by multiplying all the existing prime numbers together (including *M*) and add 1.
(3) *N* is either prime (in which case *M* is not the largest prime, QED) or it has factors.
(4) The factors of *N* cannot be any prime before *M* or *M* itself, because of the 1 we added which will appear as a remainder. Thus any factors of *N* will be larger primes than *M*, QED.

This is a surprising result as one would expect very large numbers to break up into factors.

prism a solid figure whose cross section is constant in planes drawn perpendicular to its axis. A cube, for example, is a rectangular prism with all faces (bases and sides) the same shape and size.

A cylinder is a prism with a circular cross section.

probability likelihood, or chance, that an event will occur, often expressed as odds, or numerically as a fraction or decimal. In general, the probability that *n* particular events will happen out of a total of *m* possible events is *n/m*. A certainty has a probability of 1; an impossibility has a probability of 0. Empirical probability is defined as the number of successful events divided by the total possible number of events.

prism

triangular prism

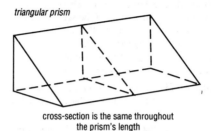

cross-section is the same throughout
the prism's length

trapezoidal prism

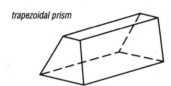

pentagonal prism

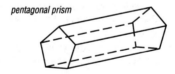

In tossing a fair coin, the chance that it will land 'heads' is the same as the chance that it will land 'tails' – that is, 1:1 or even; mathematically, this probability is expressed as 1/2 or 0.5. The odds against any chosen number coming up on the roll of a fair die are 6:1; the probability is 1/6 or 0.1666... . If two dice are rolled there are $6 \times 6 = 36$ different possible combinations. The probability of a double (two numbers the same) is 6/36 or 1/6 since there are six doubles in the 36 events: (1,1), (2,2), (3,3), (4,4), (5,5), and (6,6).

problem solving technique for solving problems efficiently, involving a special set of skills. Some of the skills are listing methods, trial and error methods, analytical methods, numerical methods, use of mathematical models, and so on.

product the result obtained from multiplying two numbers or variables. For example, ab is the product of a and b.

program a set of instructions that control the operation of a computer, usually written in a special computer language, such as Basic, Pascal, or Logo.

progression sequence of numbers each formed by a specific relationship to its predecessor. An *arithmetic progression* has numbers that increase or decrease by a common sum or difference (for example, 2, 4, 6, 8; a *geometric progression* has numbers each bearing a fixed ratio to its predecessor (for example, 3, 6, 12, 24); and a *harmonic progression* has numbers whose ◊reciprocals are in arithmetical progression, for example 1, $^1/_2$, $^1/_3$, $^1/_4$.

projectile an object ejected into the air, such as a bullet, shell, rocket, or missile. The theoretical path of a projectile is usually a parabola.

proof a set of arguments used to deduce a mathematical theorem from a set of axioms.

proper fraction or *simple fraction* or *common fraction* a fraction whose value is less than 1. The numerator has a lower value than the denominator in a proper fraction.

proportion the relation of a part to the whole (usually expressed as a fraction or percentage). For example, the proportion of children having school meals is 70%.

protractor instrument used to measure a flat ◊angle.

prove to demonstrate the truth of a proposition by reasoning. For example, to prove that a rectangle with four equal angles is a rectangle, you need to prove (1) the angles add up to 360°; (2) there are four equal angles; (3) each angle must be 360°/4 = 90°.

pyramid in geometry, a solid figure with triangular side-faces meeting at a common vertex (point) and with a ◊polygon as its base. The

protractor

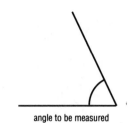

angle to be measured

angle is 64°

volume V of a pyramid is given by

$$V = {}^1\!/_3 Bh$$

where B is the area of the base and h is the perpendicular height.

Pyramids are generally classified by their bases. For example, the Egyptian pyramids have square bases, and are therefore called square

pyramid

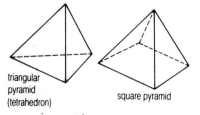

triangular
pyramid
(tetrahedron)

square pyramid

Pythagoras' theorem

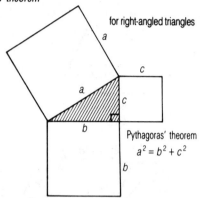

for right-angled triangles

a

c

a

c

b

Pythagoras' theorem
$$a^2 = b^2 + c^2$$

b

pyramids. Triangular pyramids are also known as tetrahedra ('four sides').

Pythagoras' theorem in geometry, a theorem stating that in a right-angled triangle, the area of the square on the hypotenuse (the longest side) is equal to the sum of the areas of the squares drawn on the other two sides. If the hypotenuse is a units long and the lengths of the other sides are b and c, then

$$a^2 = b^2 + c^2$$

quadrant one quarter of the circumference of a circle. On a graph, the first quadrant lies in the region where both x and y are positive. The remaining three quadrants are numbered in an anticlockwise direction. Thus in the second quadrant, x is negative and y is positive, in the third quadrant, both are negative, and in the fourth quadrant x is positive and y negative.

quadrant

quadrants of a graph

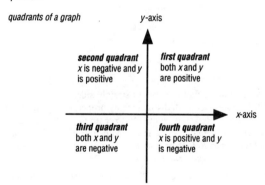

quadratic equation a polynomial equation of second degree (that is, an equation containing as its highest power the square of a variable, such as x^2). The general formula of such equations is

$$ax^2 + bx + c = 0$$

in which a, b, and c are real numbers, and only the coefficient a cannot equal 0. In ▷coordinate geometry, a quadratic function represents a

◊parabola, and the quadratic equation may be solved to find the points where the parabola intercepts the *x*-axis.

Some quadratic equations can be solved by ◊factorization or ◊completing the square, or the values of *x* can be found by using the formula for the general solution

$$x = \frac{-b \pm \sqrt{(b^2 - 4ac)}}{2a}$$

Depending on the value of the discriminant

$$b^2 - 4ac$$

a quadratic equation has two real, two equal or two complex roots (solutions). When

$$b^2 - 4ac > 0$$

there are two distinct real roots. When

$$b^2 - 4ac = 0$$

there are two equal real roots. When

$$b^2 - 4ac < 0$$

there are two distinct complex roots.

quadrilateral a ◊plane figure with four straight sides. The following are all quadrilaterals, each with distinguishing properties: ***square*** with four equal angles, four axes of symmetry; ***rectangle*** with four equal angles, two axes of symmetry; ***rhombus*** with four equal sides, two axes of symmetry; ***parallelogram*** with two pairs of parallel sides, rotational symmetry; and ***trapezium*** one pair of parallel sides.

quantity a property of an entity that can be represented by numbers.

quarter any of the four equal parts of a whole; half of a half.

quartile in statistics, any one of the three values of a variable that divide its distribution into four parts of equal frequency. They comprise the ***lower quartile*** (or 25th ◊percentile), below which lies the lowest 25% frequency distribution of a variable; the ◊***median*** (or 50th percentile), which forms the dividing line between the upper, middle 25% and the lower, middle 25%; and the ***upper quartile*** (or 75th percentile),

quadrilateral

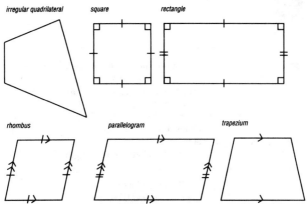

above which lies the top 25%. The difference of value between the upper and lower quartiles is known as the interquartile range which is a useful measure of the dispersion of a statistical distribution because it is not affected by freak values (see ◊range).

questionnaire a question sheet used in statistical surveys of people's opinions.

quotient the result of dividing one number or variable into another.

R

radian (symbol rad) an alternative unit to the ◊degree for measuring angles. It is the angle at the centre of a circle when the centre is joined to the two ends of an arc (part of the circumference) equal in length to the radius of the circle. There are 2π (approximately 6.284) radians in a full circle (360°). One radian is approximately 57°, and 1° is $\pi/180$ or approximately 0.0175 radians. Radians are commonly used to specify angles in ◊polar coordinates.

Radian measure gives simple formulae for the length of arc and the area of a sector of a circle:

$$\text{length of arc} = r\,\theta$$

$$\text{area of sector} = {}^1\!/_2 r^2\theta$$

where r is the length of the radius and τ is the angle measure in radians.

radius a straight line from the centre of a circle to its circumference, or from the centre to the surface of a sphere.

random event an event which is allowed to occur without any attempt being made to bias the outcome or consequence of the event. For example, if a book is opened at random, all page numbers are equally likely.

random numbers a sequence of figures in which all figures within a certain range have an equal chance of occurring at equal frequencies over a given period. Many calculators have a random-number button which can be used instead of a dice in games of chance. Computers also have a capacity to produce random numbers.

range in statistics, a measure of dispersion in a frequency distribution, equalling the difference between the largest and smallest values of the variable. The range is sensitive to freak values in the sense that it will give a distorted picture of the dispersion if one measurement is unusually large or small. The ◊interquartile range is often preferred.

range of a ◊function, the numbers on to which the base set of numbers, or ◊domain, is mapped. All the functions used in elementary mathematics have real numbers as their range.

rate of change change per unit of time. See ◊time-distance graph.

ratio measure of the relative size of two quantities or of two measurements (in similar units), expressed as a proportion. For example, the ratio of vowels to consonants in the alphabet is 5:21; the ratio of 500 m to 2 km is 500:2,000, or 1:4.

rational number any number that can be expressed as an exact fraction (with a denominator not equal to 0); that is, as $a \times b$ where a and b are integers. For example, 2, $^1/_4$, $1^5/_4$, $-^3/_5$ are all rational numbers, whereas π (which represents the constant 3.141592...) is not. Numbers such as π are called ◊irrational numbers.

The decimal form of any rational number will either terminate (for example, $^3/_5 = 0.6$) or have a repeating pattern (for example, 0.070707 ...).

real number any of the ◊rational numbers (which include the integers) or ◊irrational numbers. Real numbers exclude ◊imaginary numbers, found in ◊complex numbers of the general form $a + bi$ where i = $\sqrt{-1}$, although these do include a real component a.

reciprocal the result of dividing a given quantity into 1. Thus the reciprocal of 2 is $^1/_2$; of $^2/_3$ is $^3/_2$; of x^2 is $1/x^2$ or x^{-2}. Reciprocals are used to replace division by multiplication, since multiplying by the reciprocal of a number is the same as dividing by that number.

On a calculator, the reciprocals of all numbers except 0 are obtained by using the button marked $1/x$.

rectangle quadrilateral (four-sided plane figure) with opposite sides equal and parallel and with each interior angle a right angle (90°). Its area A is the product of the length l and width w; that is,

$$A = l \times w$$

A rectangle with all four sides equal is a ◊square.

A rectangle is a special case of a ◊parallelogram. The diagonals of a rectangle are equal and bisect each other.

rectangular axis another name for ◊Cartesian axis.

rectangular hyperbola the graph of the ◊function $y = 1/x$. The x and y axes are ◊asymptotes to this curve.

rectangular hyperbola

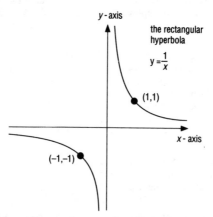

the rectangular
hyperbola

$$y = \frac{1}{x}$$

rectangular prism another name for a ◊cuboid.

recurring decimal any decimal ◊fraction that never ends but goes on repeating a number or group of numbers after the decimal point. Recurring decimals occur when fractions whose denominators are not simple multiples of 2 or 5 are converted into decimals. For example, $1/3 = 0.3333333...$, $1/6 = 0.1666666...$, $2/11 = 0.0181818...$

reduce to make smaller, as opposed to ◊enlarge.

reduce to lowest terms to cancel a fraction to its lowest common factor (the lowest figure which is an exact multiple of both numerator and denominator).

re-entrant polygon a ◊polygon that is not completely convex. It has at least one interior angle greater than 180°.

reflection

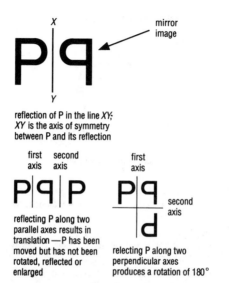

reflection of P in the line *XY*; *XY* is the axis of symmetry between P and its reflection

reflecting P along two parallel axes results in translation — P has been moved but has not been rotated, reflected or enlarged

relecting P along two perpendicular axes produces a rotation of 180°

reflection a ◊transformation that maps a shape across a line so that the line forms an axis of symmetry. Reflections in two perpendicular axes produce a rotation of 180° (a half turn). Reflections can be used to deduce many properties of shapes.

reflex angle an angle greater than 180° but less than 360°. See ◊angle for other angle types.

region a space enclosed by arcs in a ◊network (an extra region exists that is external to the network). In a closed network the numbers of regions, nodes, and arcs are related by the simple formula

$$\text{nodes} + \text{regions} = \text{arcs} + 2$$

regular of geometric figures, having all angles and sides equal. Also, of solids, having bases comprised of regular ◊polygons.

region

topological network with four regions

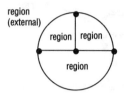

region
(external)

region | region

region

this network is closed, and so its regions,
arcs, and nodes are related by the formula:

nodes + regions = arcs + 2
(4)   (4)   (6)

relation a connection between two sets of numbers; see ◊function.
The most important relation is that of = (equivalence), which has the
rules:

(1) $a = a$ (everything is equivalent to itself)
(2) $a = b$ implies that $b = a$ (reversibility)
(3) $\{a = b$ and $b = c\}$ implies that $a = c$.

These rules are the foundation of all work in algebra.

remainder the part left over when one number cannot be exactly
divided by another. For example, the remainder of 11 divided by 3 is 2;
the remainder may be represented as a fraction or decimal. Decimal
remainders are either recurring (0.66666...), cyclic (0.37373737), or
terminating (0.125).

representative fraction (RF) the scale used in an enlargement,
expressed as a fraction. It is commonly used in maps and plans; for
example, RF 1:100 would be used for plans in which 1 cm represents
1 m.

rhombus in geometry, an equilateral (all sides equal) ◊parallelogram.
Its diagonals bisect each other at right angles, and its area is half the
product of the lengths of the two diagonals. A rhombus whose internal
angles are 90° is called a ◊square.

right angle an angle of 90°.

right-angled triangle triangle in which one of the angles is a right angle (90°). It is the basic form of triangle for defining trigonometrical ratios (for example, sine, cosine, and tangent) and for which ◊Pythagoras' theorem holds true. Its area is equal to half the product of the lengths of the two shorter sides.

Any triangle constructed with its hypotenuse (longest side) as the diameter of a circle, with its opposite vertex (corner) on the circumference, is a right-angled triangle. This is a fundamental theorem in geometry.

right-angled triangle

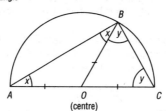

right-angled triangle

to prove that any triangle constructed so that its vertices are on the circumference of a circle and its longest side is the diameter must be a right-angled triangle

(centre)

triangles *AOB* and *BOC* are isosceles because
OA = *OB* = *OC* (radii of the circle), therefore:
angle *BAO* = angle *OBA* = *x*,
and angle *BCO* = *OBC* = *y*
(base angles of an isosceles triangle)

$2x + 2y = 180°$ (angle sum of a triangle)
$x = y = 90°$
therefore, triangle *ABC* is right-angled

Roman numerals an ancient European number system using symbols different from Arabic numerals (the ordinary numbers 1, 2, 3, 4, 5, and so on). The seven key symbols in Roman numerals, as represented today, are I (1), V (5), X (10), L (50), C (100), D (500) and M (1,000). There is no zero, and therefore no place-value as is fundamental to the Arabic system. The first ten Roman numerals are I, II, III, IV (or IIII), V, VI, VII, VIII, IX, and X. When a Roman symbol is preceded by a symbol of equal or greater value, the values of the symbols are added (XVI = 16). When a symbol is preceded by a symbol of less value, the values are subtracted (XL = 40). A horizontal bar over a symbol indicates a factor of 1,000 ($\bar{X}$ = 10,000). Although addition and subtraction are fairly straightforward using Roman numerals, the absence of a zero makes other arithmetic calculations (such as multiplication) clumsy and difficult.

root of an equation, a value that makes the equation true. For example, $x = 0$ and $x = 5$ are roots of the equation $x^2 - 5x = 0$.

root the inverse of an ◊exponent.

On a calculator, roots may be found by using the buttons marked $x^{1/y}$ or inv x^y. For example, the cubed root of twenty seven is found by pressing 27 $x^{1/y}$ 3.

root-mean-square (RMS) value obtained by taking the ◊square root of the mean of the squares of a set of values; for example the RMS value of four quantities a, b, c, and d is

$$\sqrt{\frac{(a^2 + b^2 + c^2 + d^2)}{4}}$$

rotation a ◊transformation in which a figure is turned about a given point, known as the ***centre of rotation***. A rotation of 180° is known as a half turn.

It is also useful to rotate a plane figure about an axis to produce a ***solid of revolution***; the volume of the solid can then be calculated by integration.

rounding a process by which a number is approximated to the nearest above or below with one less decimal place. For example, 34.3583

rotation

rotation of A around a centre of rotation *P*

would be rounded to 34.358, whereas 34.3587 would be rounded to 34.359. When unwanted decimals are simply left out, the process is known as *truncating*.

Rounding can produce considerable errors, especially when the rounded numbers are multiplied by other rounded numbers. For example, if the volume of a cuboid is calculated as $3 \times 4 \times 5$ when the true dimensions are $3.4 \times 4.4 \times 5.4$, the rounding error is 20.784 out of 80.784, an error of over 25%.

route a pathway on a ◊network.

row in a ◊matrix, a horizontal line of numbers. Matrices are made up of rows and columns. A matrix that consists of one row only is called a *row matrix*.

S

sample in statistics, a small set taken from a larger one in order to construct or test a theory about the whole. The statistical process consists of collecting data from samples, analysing them, and making predictions about the characteristics of the whole population. These predictions may then be tested by taking further samples.

satisfy the process in which values of variables are said to satisfy equations because they make the equation true. For example, $x^2 = 9$ is satisfied by the values 3 and –3.

scalar quantity in mathematics and science, a quantity that has magnitude but no direction, as distinct from a ◊vector quantity, which has a direction as well as a magnitude. Temperature, mass, and volume are scalar quantities.

scale the numerical relationship, expressed as a ◊ratio, between the actual size of an object and the size of an image that represents it on a map, plan, or diagram.

scale factor of an enlargement, the factor by which the original object is multiplied in order to achieve an enlarged image.

scale factor

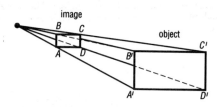

the scale factor of the enlargement of *ABCD* is 3 because

$$\frac{A'B'}{AB} = \frac{B'C'}{BC} = \frac{C'D'}{CD} = \frac{A'D'}{AD} = \frac{3}{1}$$

scatter diagram

scatter diagram relating test marks in mathematics and English in a class of 30 students

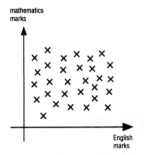

pattern suggests that there is no relationship between mathematics and English marks

scatter diagram relating the heights and weights of 30 female students

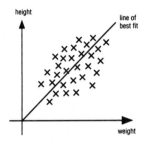

pattern suggests that there is a positive relationship between height and weight

scatter diagram or *scattergram* a diagram whose purpose is to establish whether or not a connection or ◊correlation exists between two variables, for example between life expectancy and GNP. Each observation is marked with a dot in a position that shows the value of both variables. The pattern of dots is then examined to see if they show any underlying trend by means of a *best-fit line* (a straight line drawn so that its distance from the various points is as short as possible).

secant in trigonometry, the function of a given angle in a right-angled triangle, obtained by dividing the length of the hypotenuse (the longest side) by the length of the side adjacent to the angle. It is the ◊reciprocal of the ◊cosine (secant = 1/cosine).

second one-sixtieth of a minute. It takes one second to say 'one pink elephant' briskly.

section see ◊cross section and ◊conic section.

sector part of a circle enclosed by two radii and the arc that joins them.

segment of a line, the part between two points; usually the whole line is regarded as infinite.

secant

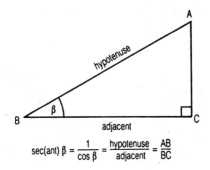

$$\sec(\text{ant}) \, \beta = \frac{1}{\cos \beta} = \frac{\text{hypotenuse}}{\text{adjacent}} = \frac{AB}{BC}$$

segment part of a circle cut off by a straight line or ◊chord, running from one point on the circumference to another. All angles in the same segment are equal.

semi or *demi* or *hemi* prefix meaning half, as in semicircle and semi-detached house.

sense the orientation of a vector. Each vector has an equivalent vector of the opposite sense. The combined effect of two vectors of opposite sense is a zero vector.

sequence set of elements, especially numbers, arranged in order according to some rule. See ◊arithmetic progression, ◊geometric progression, and ◊series.

series the sum of the terms of a sequence. Series may be convergent (the limit of the sum is a finite number) or divergent (the limit is infinite). For example:

$$1 + \frac{1}{2} + \frac{1}{4} + \frac{1}{8} + \ldots$$

is a convergent series because the limit of the sum is 2;

$$1 + \frac{1}{2} + \frac{1}{3} + \frac{1}{4} + \ldots$$

is a divergent series, as it can be made to exceed any given sum.

segment

angles in a segment

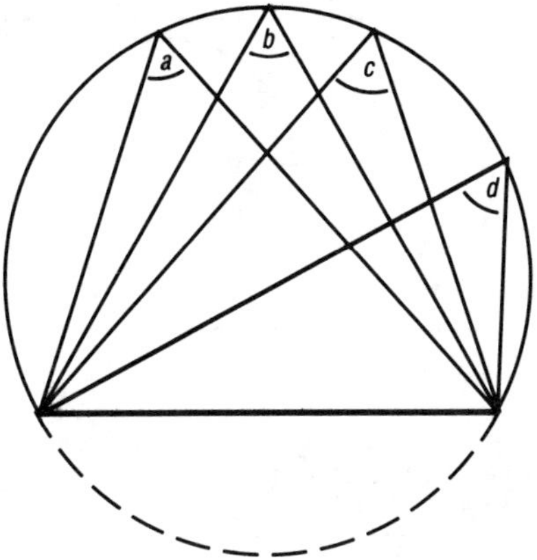

a = *b* = *c* = *d*
all such angles in a segment are equal

set or ***class*** any collection of defined things (elements), provided the elements are distinct and that there is a rule to decide whether an element is a member of a set. It is usually denoted by a capital letter and indicated by curly brackets { }.

For example, *L* may represent the set that consists of all the letters of the alphabet. The symbol $\in$ stands for 'is a member of'; thus $p \in L$ means that *p* belongs to the set consisting of all letters, and $4 \notin L$ means that 4 does not belong to the set consisting of all letters.

There are various types of sets. A ***finite set*** has a limited number of members, such as the letters of the alphabet; an ***infinite set*** has an unlimited number of members, such as all whole numbers; an ***empty*** or ***null set*** has no members, such as the number of people who have swum across the Atlantic Ocean, written as { } or $\varnothing$; a ***single-element set*** has only one member, such as days of the week beginning with M, written as {Monday}.

set

Venn diagram of two intersecting sets

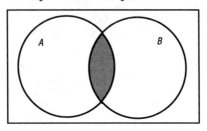

Venn diagram showing the whole numbers from 1 to 20 and the subsets of the prime and odd numbers

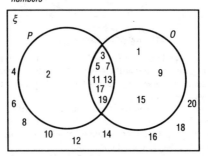

ξ = set of whole numbers from 1 to 20

O = set of odd numbers

P = set of prime numbers

the intersection of P and O $(P \cap O)$ contains all the prime numbers that are also odd

Equal sets have the same members; for example, if $W = \{$days of the week$\}$ and $S = \{$Sunday, Monday, Tuesday, Wednesday, Thursday, Friday, Saturday$\}$, it can be said that $W = S$. Sets with the same number of members are *equivalent sets*. Sets with some members in common are *intersecting sets*; for example, if $R = \{$red playing cards$\}$ and $F = \{$face cards$\}$, then R and F share the members that are red face cards. Sets with no members in common are *disjoint sets*, such as $\{$minerals$\}$ and $\{$vegetables$\}$. Sets contained within others are *subsets*; for example, $\{$vowels$\}$ is a subset of $\{$letters of the alphabet$\}$.

Sets and their interrelationships are often illustrated by a ◊Venn diagram.

shape the form of an object, defined by its outline. Plane shapes are two-dimensional; solid shapes are three-dimensional.

sign symbol that indicates whether a number is positive or negative (+ or –), the operation that is to be carried out on a number of set of numbers (for example, $+$, $\times$, $\div$, and $\sqrt{}$), or the relationship that exists between two numbers or sets of numbers (for example, $=$, $\neq$, $>$, and $<$).

significant figures the figures in a number that, by virtue of their place value, express the magnitude of that number to a specified degree of accuracy. The final significant figure is rounded up if the following digit is greater than 5. For example, 5,463,254 to three significant figures is 5,460,000; 3.462891 to four significant figures is 3.463; 0.00347 to two significant figures is 0.0035.

similar of plane or solid figures, having the same shape but a different size or orientation. Two similar figures will have the same corresponding angles, and the ratio of the lengths of their corresponding sides or edges will be constant. For example, a rectangle of length 8 cm and height 6 cm and a larger rectangle of length 12 cm and height 9 cm are similar because the ratio of their lengths (8:12 or 1:1.5) and the ratio of their heights (6:9 or 1:1.5) are equal. Similar figures may be produced by enlargement.

The ratio of the areas of similar plane shapes are equal to the square of the ratio of their corresponding sides. For example, the ratio of the areas of the two rectangles described above is 48:108, which can be reduced to 1:2.25 or 1:(1.5^2).

similar

similar figures

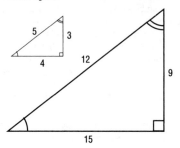

these triangles are similar because:
(1) their corresponding angles are equal
(2) the lengths of their corresponding sides are
in the same ratio —each of the sides of the
larger triangle is three times longer than the
corresponding side of the smaller triangle

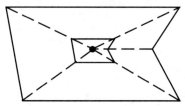

an enlargement produces a similar figure

simplify of a fraction, to reduce to lowest terms (to cancel to the low-est common factor for both numerator and denominator). Also, in alge-bra, to condense an algebraic expression by grouping similar terms and reducing constants to their lowest terms. For example, the expression

$$a + 2b + b + 2a - 2(a + b)$$

can be simplified to $a + b$.

simultaneous equations one of two or more algebraic equations that contain two or more unknown quantities that may have a unique solution. For example, in the case of two linear equations with two unknown variables, such as

$$\text{(i) } x + 3y = 6$$

and

$$\text{(ii) } 3y - 2x = 4$$

the solution will be those unique values of x and y that are valid for both equations. Linear simultaneous equations can be solved by using algebraic manipulation to eliminate one of the variables, ◊coordinate geometry, or matrices (see ◊matrix).

For example, by using algebra, both sides of equation (i) could be multiplied by 2, which gives

$$2x + 6y = 12$$

This can be added to equation (ii) to get

$$9y = 16$$

which is easily solved:

$$y = {}^{16}/_9$$

The variable x can now be found by inserting the known y value into either original equation and solving for x. Another method is by plotting the equations on a graph, because the two equations represent straight lines in ◊coordinate geometry and the coordinates of their point of intersection are the values of x and y that are true for both of them (see ◊graphical methods).

A third method of solving linear simultaneous equations involves manipulating matrices. If the equations represent either two parallel lines or the same line, then there will be no solutions or an infinity of solutions respectively.

sine in trigonometry, a function of an angle in a right-angled triangle found by dividing the length of the side opposite the angle by the length of the hypotenuse (the longest side). Sine is usually shortened to *sin*.

sine rule in trigonometry, a rule that relates the sides and angles of a triangle, stating that the ratio of the length of each side and the sine of

sine

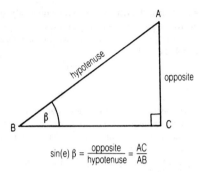

$$\sin(e)\ \beta = \frac{\text{opposite}}{\text{hypotenuse}} = \frac{AC}{AB}$$

the angle opposite is constant. If the sides of a triangle are a, b, and c, and the angles opposite are A, B, and C, respectively, then the sine rule may be expressed as:

$$\frac{a}{\sin A} = \frac{b}{\sin B} = \frac{c}{\sin C}$$

sine rule

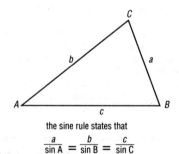

the sine rule states that

$$\frac{a}{\sin A} = \frac{b}{\sin B} = \frac{c}{\sin C}$$

or

$$\frac{\sin A}{a} = \frac{\sin \hat{B}}{b} = \frac{\sin \hat{C}}{c}$$

Each of the expressions $a/\sin A$ is equal to twice the radius of the circumcircle of the triangle.

sketch a rough drawing or graph in which the main features are marked clearly.

skew distribution in statistics, a distribution in which frequencies are not balanced about the mean. For example, low wages are earned by a great number of people, while high wages are earned by very few. However, because the high wages can be very high they pull the average up the scale making the average wage look higher than it really is.

skew distribution

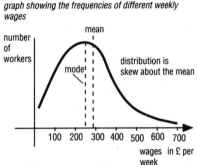

graph showing the frequencies of different weekly wages

skew lines straight lines that are not parallel and yet do not meet since they lie in a different plane. Every pair of skew lines has a minimum distance between them, which is the length of their common perpendicular.

slide rule mathematical instrument with pairs of logarithmic sliding scales, used for rapid calculations, including multiplication, division, and the extraction of square roots. It has been largely superseded by the electronic calculator.

slope another name for ◊gradient.

skew lines

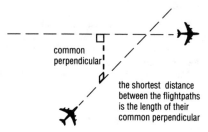

the flight paths of two different jets flying at different
heights and in different directions are skew lines

common perpendicular

the shortest distance
between the flightpaths
is the length of their
common perpendicular

solution in algebra, the value of a variable that satisfies a given equation; see ◊root.

solution set the set of values that satisfies an ◊inequality relationship; for example, the inequality relation

$$3 > n > 10$$

(where n is an integer) has the solution set [4, 5, 6, 7, 8, 9], while

$$3 \leqslant n \leqslant 10$$

has [3, 4, 5, 6, 7, 8, 9, 10] as its solution set.

solve to find the ◊roots of an equation or the answer to a problem.

speed the rate at which an object moves. The constant speed v of an object may be calculated by dividing the distance s it has travelled by the time t taken to do so, and may be expressed as:

$$v = \frac{s}{t}$$

The usual units of speed are metres per second or kilometres per hour.

Speed is a scalar quantity in which direction of motion is unimportant (unlike the vector quantity ◊velocity, in which both magnitude and direction must be taken into consideration).

speed-time graph graph used to describe the motion of a body by illustrating how its speed or velocity changes with time. The gradient of the graph gives the object's acceleration: if the gradient is zero (the graph is horizontal) then the body is moving with constant speed or uniform velocity; if the gradient is constant, the body is moving with uniform acceleration. The area under the graph gives the total distance travelled by the body.

sphere a perfectly round solid with all points on its surface the same distance from the centre. This distance is the radius of the sphere. For a sphere of radius r, the volume

$$V = \frac{4}{3}\pi r^3$$

and the surface area

$$A = 4\pi r^2$$

The Earth is a sphere that is slightly flattened at its north and south poles.

spiral a plane curve formed by a point winding round a fixed point from which it distances itself at regular intervals, for example the spiral traced by a flat coil of rope. Various kinds of spirals can be generated mathematically – for example, an equiangular or logarithmic spiral (in which a tangent at any point on the curve always makes the same angle with it) and an ◊involute. Spirals also occur in nature as a normal consequence of accelerating growth, such as the spiral shape of the shells of snails and some other molluscs.

spreadsheet in computing, a program that mimics a sheet of ruled paper, divided into columns and rows. The user enters values in the sheet, then instructs the program to perform some operation on them, such as totalling a column or finding the average of a series of numbers. Highly complex numerical analyses may be built up from these simple steps.

Spreadsheets are widely used in business for forecasting and financial control. The first spreadsheet program, VisiCalc, appeared in 1979. The best known include Lotus 1–2–3 and Excel.

square in geometry, a quadrilateral (four-sided) plane figure with all sides equal and each angle a right angle. Its diagonals bisect each other at right angles. The area A of a square is the length l of one side multiplied by itself

$$A = l \times l$$

Also, any quantity multiplied by itself is also termed a square, represented by an ◊exponent of power 2; for example:

$$4 \times 4 = 4^2 = 16$$

and

$$6.8 \times 6.8 = 6.8^2 = 46.24$$

An algebraic term is squared by doubling its exponent and squaring its coefficient if it has one; for example:

$$(x^2)^2 = x^4$$

and

$$(6y^3)^2 = 36y^6$$

A number that has a whole number as its ◊square root is known as a **perfect square**; for example, 25, 144 and 54,756 are perfect squares (with roots of 5, 12, and 234, respectively).

The expansion of the sums and differences of squares is important:

$$(a + b)^2 = a^2 + b^2 + 2\,ab$$
$$(a - b)^2 = a^2 + b^2 - 2ab$$

Also the factorization of the difference of squares is worth remembering:

$$x^2 - y^2 = (x - y)(x + y)$$

Note that $a^2 + b^2$ does not normally factorize.

square root a number that when squared (multiplied by itself) equals a given number. For example, the square root of 25 (written $\sqrt{25}$) is ± 5, because

$$5 \times 5 = 25, \text{ and } (-5) \times (-5) = 25$$

As an ◊exponent, a square root is represented by $^1/_2$ – for example, $16^{1/2} = 4$. This may be justified by the rule of indices:

$$a^{1/2} \times a^{1/2} = a^1 = a$$

standard measure against which others are compared. For example, until 1960 the standard for the metre was the distance between two lines engraved on a platinum–iridium bar kept at the International Bureau of Weights and Measures in Sèvres, France.

standard deviation in statistics, a measure (s) of the spread of data. The deviation (difference) of each data item from the ◊mean is found, and their values squared. The mean value of these squares is then calculated. The standard deviation is the square root of this mean.

If n is the number of items of data, x is the value of each item, and $\bar{x}$ is the mean value, the standard deviation (s) may be given by the formula

$$s = \sqrt{\frac{\sum(x_i - \bar{x})^2}{n}}$$

where $\sum$ indicates that the squared differences between the value of each item of data and the mean should be summed.

For example, to find the standard deviation of the ages of a group of eight people in a room, the mean is first found (in this case by adding all the ages together and dividing the total by 8), and the deviations between all the individual ages and the mean calculated. Thus, if the ages of the eight people are 14, 14.5, 15, 15.5, 16, 17, 19, and 21, the mean age is $132 \div 8 = 16.5$. The deviations between the individual ages and this mean age are –2.5, –2.0, –1.5, –1.0, –0.5, +0.5, +2.5 and +4.5. These values are then squared to give 6.25, 4.00, 2.25, 1.00, 0.25, 0.25, 6.25, and 20.25, with a mean value of $40.5 \div 8 = 5.0625$. The square root of this figure is 2.25, which is the standard deviation in years.

standard form or *scientific notation* a method of writing numbers often used by scientists, particularly for very large or very small numbers. The numbers are written in the form $a \times 10^n$, where a is a number greater than 1 and less than 10, and n is a positive or negative integer. For example, 3,950,000 becomes 3.95×10^6. The speed of light expressed in standard form is 2.9979×10^8 metres.

statistics the branch of mathematics concerned with the collection and interpretation of data. For example, to determine the ◊mean age of the children in a school, a statistically acceptable answer might be obtained by calculating an average based on the ages of a representative sample, consisting, for example, of a random tenth of the pupils from each class. ◊Probability is the branch of statistics dealing with predictions of events.

Statistics has many applications in government, business, industry, and commerce.

straight line a line that does not bend or curve. The graph of a linear relationship is a straight line and is often presented in the form $y = mx + c$, where m is the slope, or gradient, of the line and c is the y-intercept (the point at which the line cuts the y-axis).

subject the term in a formula that is explicitly found by substituting values for the other variables. For example, S is the subject of the formula

$$S = \frac{4\pi r^3}{3}$$

Often formulae have to be arranged so that a different variable becomes the subject. This is done by treating the formula as an equation and applying the following rules:

(1) add the same quantity (+ or −) to both sides of the equation;
(2) multiply both sides of the equation by the same quantity;
(3) take exponents of both sides of the equation.

For example, to make r the subject of the equation above:

$$S = \frac{4\pi r^3}{3}$$

multiply both sides of the equation by 3

$$3S = 4\pi r^3$$

multiply both sides of the equation by $1/4\pi$

$$\frac{3S}{4\pi} = r^3$$

take the cube root of both sides of the equation

$$\sqrt[3]{\frac{3S}{4\pi}} = r$$

subset

Venn diagram showing B as a subset of A

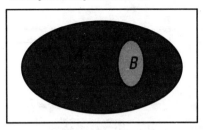

subset set drawn from a larger set. For example, the girls in a class make up a subset if the class contains both boys and girls.

substitute to put values in the place of variables in an algebraic expression or formula. An algebraic expansion or simplification can be checked by substituting simple values for each variable. For example, to check that

$$x^3 + y^3 = (x + y)(x^2 + y^2 - xy)$$

the value $x = 1$ and $y = 2$ might be substituted in both sides of the expression, giving left-hand side:

$$1^3 + 2^3 = 1 + 8 = 9$$

right-hand side:

$$(1 + 2)(1^2 + 2^2 - 2) = 3 \times 3 = 9$$

The two sides are the same, so the expansion of $x^3 + y^3$ is correct.

subtraction taking one number or quantity away from another, one of the four basic operations of arithmetic. Subtraction is not a ◊commutative operation because:

$$a - b \neq b - a$$

nor is it an ◊associative operation because:

$$a - (b - c) \neq (a - b) - c)$$

surface area

surface areas of common three-dimensional shapes

surface area of a **cube**
(faces are identical)
= 6 × area of each face
= 6 l^2

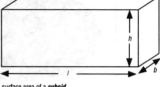

surface area of a **cuboid**
(opposite faces are identical)
= area of two end faces = area of two sides
  + area of top and base
= 2 lh + 2 hb + 2 lb
= 2(lh + hb × lb)

surface area of a **cylinder**
= area of a curved surface + area of top
and base
= 2π × (radius of cross-section × height)
  + 2π (radius of cross-section)^2
= $2\pi rh + 2\pi r^2$
= $2\pi r (h + r)$

surface area of a **cone**
= area of a curved surface + area of base
= π × (radius of cross-section × slant height)
  + π (radius of cross-section)^2
= $\pi r l + \pi r^2$
= $\pi r (l + r)$

surface area of a **sphere**
= 4π × radius^2
= $4\pi r^2$

For example:

$$8 - 5 \neq 5 - 8$$
$$7 - (4 - 3) \neq (7 - 4) - 3$$

surd an expression containing the root of an ◊irrational number that can never be exactly expressed – for example, $\sqrt{3} = 1.732050808...$.

surface area the area of the outer surface of a three-dimensional shape, or solid.

survey in statistics, a method of collecting data in which people are asked to answer a number of questions (usually in the form of a questionnaire). An opinion poll is a survey. The reliability of a survey's results depends on whether the sample from which the information has been collected is free from bias and sufficiently large.

symmetry

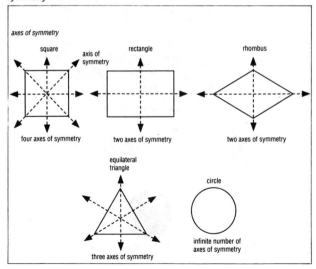

axes of symmetry

square   axis of symmetry   rectangle   rhombus

four axes of symmetry   two axes of symmetry   two axes of symmetry

equilateral triangle

circle

infinite number of axes of symmetry

three axes of symmetry

symmetry

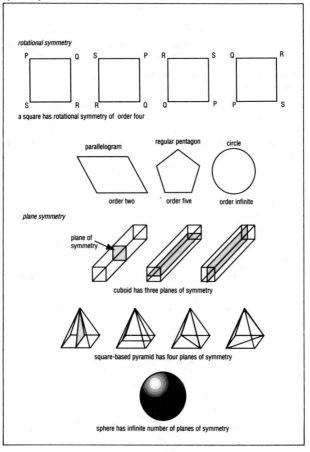

rotational symmetry

a square has rotational symmetry of order four

parallelogram — order two

regular pentagon — order five

circle — order infinite

plane symmetry

plane of symmetry

cuboid has three planes of symmetry

square-based pyramid has four planes of symmetry

sphere has infinite number of planes of symmetry

symbol a letter, figure, or sign used to represent a word or sentence. For example, $y = x^2$ states in symbols that one quantity is equal to the square of another.

symmetry exact likeness in shape about a given line (axis), point, or plane. A figure has symmetry if one half can be rotated or reflected onto the other.

Symmetry and balance are closely related in nature; this accounts for the symmetry of the human body for example.

T

table collection of information, usually numbers, arranged so that information is easy to find. *Two-way tables* are designed to be read from the top and from the side, for example a ◊multiplication table, which gives all the products of numbers from one to ten.

tabulate to organize numbers or data in a table.

tangent in geometry, a straight line that touches a curve and has the same ◊gradient as the curve at the point of contact. At a maximum or minimum, the tangent to a curve has zero gradient. Also, in trigonometry, a function of an acute angle in a right-angled triangle, defined as the ratio of the length of the side opposite the angle to the length of the side adjacent to it; a way of expressing the slope of a line.

template a specially cut piece of card, metal, or perspex used to facilitate the drawing of shapes.

terminating decimal decimal fraction with a finite number of digits (a ◊recurring decimal, by contrast, has an infinite number of digits). Only those fractions with denominators that are multiples of two and five can be converted into terminating decimals.

tessellation tiling pattern that covers space without leaving any gaps. The shapes used are not always ◊regular.

tetrahedron (plural *tetrahedra*) in geometry, a solid figure (◊polyhedron) with four triangular faces; that is, a ◊pyramid on a triangular base. A regular tetrahedron has equilateral triangles as its faces.

theorem mathematical proposition that can be deduced by logic from a set of axioms (basic facts that are taken to be true without proof). Advanced mathematics consists almost entirely of theorems and proofs, but even at a simple level theorems are important.

thousand ten hundreds, 1,000 or 10^3.

tangent

tangent to a curve

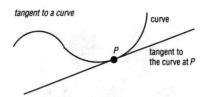

curve

tangent to
the curve at *P*

P

tangent to a circle

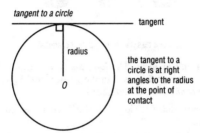

tangent

radius

O

the tangent to a
circle is at right
angles to the radius
at the point of
contact

tangent of an angle

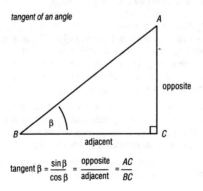

A

opposite

β

B   adjacent   *C*

$$\text{tangent } \beta = \frac{\sin \beta}{\cos \beta} = \frac{\text{opposite}}{\text{adjacent}} = \frac{AC}{BC}$$

tessellation

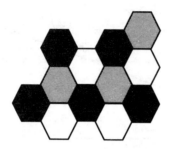

part of a tessellation of regular hexagons

ton imperial unit of mass. The *long ton*, used in the UK, is 1,016 kg/2,240 lb; the *short ton*, used in the USA, is 907 kg/2,000 lb. The *metric ton* or *tonne* is 1,000 kg/2,205 lb.

tonne the metric ton of 1,000 kg/2,204.6 lb; equivalent to 0.9842 of an imperial ◊ton.

topology branch of geometry that deals with those properties of a figure that remain unchanged even when the figure is transformed (bent, stretched) – for example, when a square painted on a rubber sheet is deformed by distorting the sheet. Topology has scientific applications, as in the study of turbulence in flowing fluids. The map of the London Underground system is an example of the topological representation of a network; connectivity (the way the lines join together) is preserved, but shape and size are not. The ◊Königsberg bridge problem is a famous problem that was solved by topology.

torus a shape like a like a doughnut with a hole in the middle. If a cylindrical pipe is bent into a circle and the two ends joined, then a torus is formed. The shape is interesting in that a closed curve can be drawn on a torus without separating it into two parts.

transformation a mapping or ◊function, especially one that causes a change of shape or position in a geometric figure. ◊Reflection,

topology

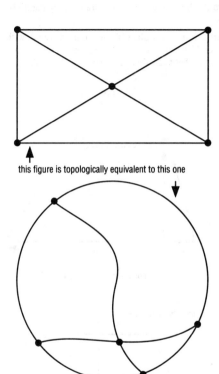

this figure is topologically equivalent to this one

◊rotation, ◊enlargement, and ◊translation are the main geometrical transformations.

translation a ◊transformation in which a figure is moved or slid to another position without turning, that is, its new axes remain parallel to the old.

transversal a line cutting two or more (usually parallel) lines in the same plane.

trapezium a four-sided plane figure (quadrilateral) with two of its sides parallel. If the parallel sides have lengths a and b and the perpendicular distance between them is h (the height of the trapezium), its area

$$A = {}^1\!/_2 h(a + b)$$

trapezium

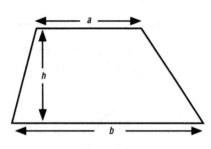

area of a trapezium $= \frac{1}{2}\ (a+b)\ \times h$

isosceles trapezium

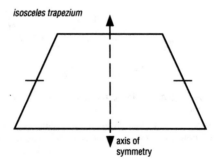

axis of
symmetry

An isosceles trapezium has its sloping sides equal, and is symmetrical about a line drawn through the midpoints of its parallel sides.

travel graph a type of ◊distance-time graph which shows the stages of a journey. Stops on the journey are shown as horizontal lines. Travel graphs can be used to find out where vehicles will overtake each other or meet if travelling in opposite directions.

tree diagram a diagram consisting only of arcs and nodes (but not loops), which is used to establish probabilities.

tree diagram

tree diagram used to work out the probability of each combination of results (winning or losing scores) on throwing a die three times

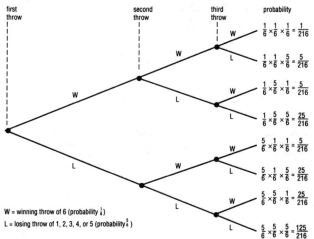

W = winning throw of 6 (probability $\frac{1}{6}$)
L = losing throw of 1, 2, 3, 4, or 5 (probability $\frac{5}{6}$)

trial in a probability experiment, each experiment or observation is termed a trial.

trial and improvement method a method of solving problems that involves making a first attempt and using the information that this yields to make a better second attempt. The problem solver does not expect to solve the problem outright but only to get nearer to a solution.

triangle in geometry, a three-sided plane figure, the sum of whose interior angles is 180°. Triangles can be classified by the relative

triangle

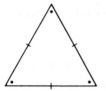

Equilateral triangle: all the sides are the same length; all the angles are equal to 60°

Isosceles triangle: two sides and two angles are the same

Scalene triangle: all the sides and angles are different

Acute-angle triangle: each angle is acute (less than 90°)

Obtuse-angle triangle: one angle is obtuse (more than 90°)

A *right-angle triangle* has one angle of 90°, the *hypotenuse* is the side opposite the right angle

Area of triangle = ½lh

Triangles are *congruent* if corresponding sides and corresponding angles are equal

Similar triangles have corresponding angles that are equal; they therefore have the same shape

lengths of their sides. A *scalene triangle* has no sides of equal length; an *isosceles triangle* has at least two equal sides; an *equilateral triangle* has three equal sides (and three equal angles of 60°).

A right-angled triangle has one angle of 90°. If the length of one side of a triangle is l and the perpendicular distance from that side to the opposite corner is h (the height or altitude of the triangle), its area

$$A = \frac{1}{2} l \times h.$$

triangular number any number that will form an equilateral triangle when arranged in dots.

trigonometry branch of mathematics that solves problems relating to plane and spherical triangles. Its principles are based on the fixed

trigonometry

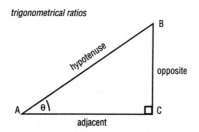

trigonometrical ratios

for any right-angled triangle with angle θ as shown the **trigonometrical ratios** are

$$\sin\theta = \frac{BC}{AB} = \frac{\text{opposite}}{\text{hypotenuse}}$$

$$\cos\theta = \frac{AC}{AB} = \frac{\text{adjacent}}{\text{hypotenuse}}$$

$$\tan\theta = \frac{BC}{AC} = \frac{\text{opposite}}{\text{adjacent}}$$

proportions of sides for a particular angle in a right-angled triangle, the simplest of which are known as the ◊sine, ◊cosine, and ◊tangent (so-called **trigonometrical ratios**). It is of practical importance in navigation, surveying, and simple harmonic motion in physics.

Using trigonometry, it is possible to calculate the lengths of the sides and the sizes of the angles of a right-angled triangle as long as one angle and the length of one side are known, or the lengths of two sides.

union a set which is formed by joining up two or more other sets.

union of sets

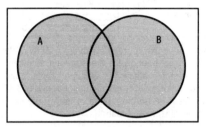

shaded area represents the union of sets A and B
A ∪ B

unit standard quantity in relation to which other quantities are measured. There have been many systems of units. Some ancient units, such as the day, the foot, and the pound, are still in use.

unit matrix the matrix which is the ◊identity for multiplication in a family of matrices. For example:

$$\begin{bmatrix} 1 & 0 \\ 0 & 1 \end{bmatrix}$$

is the unit matrix for 2 by 2 matrices.

universal set the set from which subsets are taken, the set of all the objects under consideration in a problem. The odd numbers 1, 3, 5, 7 ... and the square numbers 1, 4, 9, 16 ... are taken from the universal set of natural numbers.

universal set

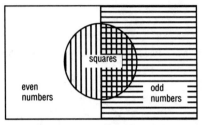

ξ = universal set of all whole numbers

upper bound a value that is greater than or equal to all the values of a given set. The upper bound of a measurement is taken as the top extreme of the possible values. For example, for a length given as 3.4 cm correct to one decimal place, the upper bound is 3.45 correct to two decimal places. The limiting value of an infinite set is the *least upper bound*. See also ◊approximation.

V

value a number or other fixed quantity applied to a ◊variable. The value of a fraction is found by dividing numerator by denominator. The value of an expression will depend on the numbers which are substituted for the variables. For example, $x^2 + y^2$ has the value 25 when $x = 3$ and $y = 4$.

variable a changing quantity (one that can take various values), as opposed to a ◊constant. For example, in the algebraic expression $y = 4x^3 + 2$, the variables are x and y, whereas 4 and 2 are constants.

A variable may be dependent or independent. Thus if y is a ◊function of x, written $y = f(x)$, such that $y = 4x^3 + 2$, the domain of the function includes all values of the *independent variable* x while the range (or codomain) of the function is defined by the values of the *dependent variable* y.

variation a practical relationship between two variables. *Direct variation* (when the values of the variables maintain a constant ratio) corresponds to

$$y = kx$$

on a straight line graph, for example distance travelled at a steady speed. *Inverse variation* (when an increase in the value of one variable results in a decrease in that of the other) corresponds to

$$y = \frac{k}{x}$$

on a rectangular hyperbolic graph, for example the price of an article and the quantity that can be bought for a fixed sum of money. In problems of direct and inverse variation, the first step is to find the value of k.

vector quantity any physical quantity that has both magnitude and direction (such as the velocity or acceleration of an object) as distinct from ◊scalar quantity (such as speed, density, or mass), which has mag-

vector

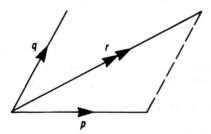

nitude but no direction. A vector is either represented geometrically by an arrow whose length corresponds to its magnitude and points in an appropriate direction, or by a pair of numbers written vertically and placed within brackets

$$\begin{bmatrix} x \\ y \end{bmatrix}$$

Vectors can be added graphically by constructing a triangle of vectors (such as the triangle of forces commonly employed in physics and engineering).

If two forces p and q are acting on a body at A, then the parallelogram of forces is drawn to determine the resultant force and direction r. p, q, and r are vectors. In technical writing, a vector is denoted by **bold** type, underlined $\underline{AB}$, or overlined $\overrightarrow{AB}$.

velocity speed of an object in a given direction. Velocity is a ◊vector quantity, since its direction is important as well as its magnitude (or speed).

The velocity at any instant of a particle travelling in a curved path is in the direction of the tangent to the path at the instant considered.

Venn diagram a diagram representing a ◊set or sets and the logical relationships between them. The sets are drawn as circles. An area of overlap between two circles (sets) contains elements that are common to both sets, and thus represents a third set. Circles

vertex

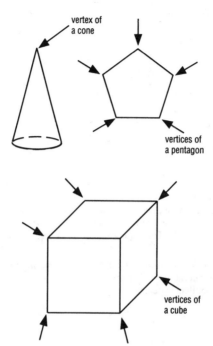

vertex of
a cone

vertices of
a pentagon

vertices of
a cube

that do not overlap represent sets with no elements in common (disjoint sets).

vertex plural *vertices* in geometry, a point shared by three or more sides of a solid figure; the point farthest from a figure's base; or the point of intersection of two sides of a plane figure or the two rays of an angle.

vertical at right angles to the horizontal plane.

A string with a heavy weight tied to it hangs in a vertical line. Such a string is called a plumb line.

vertically opposite angles

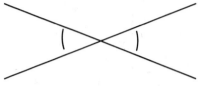

vertically opposite angles are equal

vertically opposite angles a pair of angles that lie vertically opposite each other on the same side of a transveral (a line that intersects two or more lines).

volume in geometry, the space occupied by a three-dimensional solid object. A prism (such as a cube) or a cylinder has a volume equal to the area of the base multiplied by the height. For a pyramid or cone, the volume is equal to one-third of the area of the base multiplied by the perpendicular height. The volume of a sphere is equal to $4/3\pi r^3$, where r is the radius.

vulgar fraction or *common fraction* or *simple fraction* a fraction comprising natural or whole numbers and written as a ratio, for example *a/b*, rather than in decimal form.

volume

volumes of common three-dimensional shapes

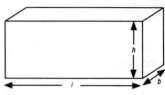

volume of a cube

= length^3

= $\bullet$

volume of a cuboid

= length × breadth × height

= $l \times b \times h$

volume of a cylinder

= π × (radius of cross-section)^2 × height

= $\pi r^2 h$

volume of a cone

= $\frac{1}{3}$ π × (radius of cross-section)^2 × height

= $\frac{1}{2} \pi r^2 h$

volume of a sphere

= $\frac{4}{3}$ π radius^3

= $\frac{4}{3} \pi r^3$

W

weigh to measure the mass of an object.

weight the force exerted on an object by ◊gravity. The weight of an object depends on its mass – the amount of material in it – and the strength of the Earth's gravitational pull, which decreases with height. Consequently, an object weighs less at the top of a mountain than at sea level. On the Moon, an object has only one-sixth of its weight on Earth, because the pull of the Moon's gravity is one-sixth that of the Earth.

whole number an ◊integer.

width the measurement from side to side of an object.

Y

yard imperial unit (symbol yd) of length, equivalent to three feet (0.9144 m).

year unit of time measurement, based on the orbital period of the Earth around the Sun.

Z

zero the number (written 0) that results when any number is subtracted from itself, or when any number is added to its negative. The product of any number with zero is itself zero. This can be proved from the definition of zero as a number which, added to others, leaves them unchanged.

The Hutchinson
Pocket Encyclopedia
1994 edition

The Hutchinson Pocket Encyclopedia packs as much information into its 640 pages as other volumes costing half as much again. Written in clear, jargon-free language, it covers all areas of knowledge from science and engineering to biographies of major sporting, cultural, and political figures, to newly updated information on all the countries of the world.

With more than 10,000 up-to-date entries, 150 maps and diagrams, and a host of useful chronologies and tables, the Pocket is quite simply the most comprehensive compact encyclopedia available.

ISBN 0 09 178286 4

Hutchinson
Pocket Dictionary
of Computing

If you use or are considering buying a personal computer, this dictionary is for you. In a compact format, here are all the terms that confuse, obscure, and mystify, made clear and intelligible to the non-expert. Included are all the latest terms – image compression, local bus, MPC – as well as basic topics such as boot disc and graphics card. Whether you are a newcomer to computing or a seasoned hand, you will find this dictionary an essential aid.

published October 1993

ISBN 0 09 178104 3

Hutchinson
Pocket Chronology
of World Events

From early humans in Asia (800,000 BC) and the construction of the Pyramids (2500 BC) to the breakup of the USSR (Dec 1991) and the 1992 Earth Summit, the Pocket Chronology lists the major events in world history, including science, politics, religion, and society. Here is the whole of world history at a glance.

published October 1993

ISBN 0 09 178275 9

Hutchinson
Pocket Dictionary
of Quotations

From Aristotle to Yeltsin, this dictionary of more than 3,000 quotations is fully indexed to enable quotations to be found from individual words within the quotation. Historical, literary, political, topical, and humorous quotations are all included, resulting in a book that can be browsed through for hours.

published October 1993

ISBN 0 09 178281 3

Hutchinson
Pocket Quiz Book

The Pocket Quiz Book enables you to pitch your knowledge against the thousands of facts in *The Hutchinson Encyclopedia*. It consists of 100 quizzes ranged in three levels of difficulty, a total of over 1,000 questions, all in a handy format that fits into a pocket. Every answer refers you to the corresponding entry in *The Hutchinson Encyclopedia* where explanations and additional material can be found.

published October 1993

ISBN 0 09 178101 9